SHOOT FOR THE MOON

Also by Richard Wiseman

RICHARD WISEMAN

SHOOT FOR THE MOON

Achieve the impossible
with the Apollo Mindset

First published in Great Britain in 2019 by

Quercus Editions Ltd
Carmelite House
50 Victoria Embankment
London EC4Y 0DZ

An Hachette UK company

A CIP catalogue record for this book is available
from the British Library

HB ISBN 978 1 78747 443 7
TPB ISBN 978 1 78747 444 4
Ebook ISBN 978 1 78747 446 8

10 9 8 7 6 5 4 3 2 1

Author photo by Brian Fischbacher
Typeset by CC Book Production
Printed and bound in Great Britain by Clays Ltd, Elcograf S.p.A.

This is dedicated to the Moon.

Thank you for helping us to look up, think big and travel far.

'We choose to go to the Moon in this decade, not because that will be easy, but because it will be hard – because that goal will serve to organize and measure the best of our energies and skills, because that challenge is one that we are willing to accept, one we are unwilling to postpone, and one which we intend to win.'

– President John F. Kennedy, September 1962[1]

CONTENTS

TIMELINE

October 1957

Soviets launch Sputnik

November 1957

Sputnik 2 carries Laika the dog into space

April 1961

Russian cosmonaut Yuri Gagarin orbits the Earth

May 1961

Alan Shepard becomes the first American in space

May 1961

Kennedy announces to Congress that America will put
a man on the Moon before the decade is out

September 1962

Kennedy delivers his 'We choose to go to the Moon'
speech at Rice University

January 1967

Tragic Apollo 1 fire results in the death of three astronauts

December 1968

Apollo 8 becomes the first manned mission to the Moon

16 July 1969

Apollo 11 astronauts Neil Armstrong, Buzz Aldrin and
Michael Collins begin their historic journey to the Moon

21 July 1969

Armstrong sets foot on the lunar surface, declaring:
'That's one small step for a man, one giant leap for mankind'

24 July 1969

Apollo 11 splashes down safely in the Pacific Ocean
and fulfils Kennedy's astonishing aim of having a man
walk on the lunar surface, and return safely to the Earth,
before the end of the decade

LIFT-OFF

Open almost any book about how to achieve your aims and ambitions and you will quickly encounter stories of genetically gifted Olympians, hard-headed CEOs and risk-taking entrepreneurs. This book presents a radically different perspective on success. It's based around a group of young and surprisingly ordinary people. Their inspirational story is little known and yet they played a central role in humanity's greatest achievement. Perhaps most important of all, once you understand how they did what they did, you can follow in their footsteps and achieve the extraordinary in your own life.

As a psychologist, I have spent much of my career examining why some people and organizations are especially successful. A few years ago, I became interested in the Moon landings, and was astonished to discover that although the technology used during the missions has been extremely well documented, very little has been written about the psychology that lies behind this remarkable achievement.[2] As I started to dig deeper, I encountered several other big surprises.

On 21 July 1969, Apollo astronaut Neil Armstrong gently set foot on the powdery surface of the Moon. Live images of this historic event were beamed back to the Earth, where over 500 million people watched in wonder. Just eight years before, President John F. Kennedy had appeared in front of Congress and famously declared that before the decade was out, America would put a man on the Moon. Looking back, it's hard to fully appreciate the enormity of the goal.

When the President made his historic announcement, America had only managed to send one astronaut on a fifteen-minute, 'up-and-down' suborbital flight. Kennedy's bold vision required several astronauts to journey over 380,000 kilometres across space, land on a distant and hostile world and safely make their way back home. Even modern-day space travel pales in comparison, with the Space Shuttle and International Space Station travelling only around 400 kilometres above the surface of the Earth. Moreover, Kennedy's goal had to be achieved within a few years, and at a time when the latest cutting-edge technology consisted of slide-rules and mainframe computers with less processing power than a modern-day smartphone.

To many, Kennedy's dream appeared to be truly impossible. Nevertheless, hundreds of thousands of people came together and attempted to transform that dream into reality. They invented new forms of technology, overcame setbacks and tragedies, and built spacecraft containing millions of bespoke parts. Against astonishing odds, the Moon landings were a spectacular success, and provided the world with an unparalleled sense of optimism and hope.

For most of us, mention of the Apollo landings conjures up images of white-suited astronauts carefully making their way across the lunar surface. While these brave and heroic individuals are obviously essential to the success of Apollo, they are far from the full story. Watch any documentary about the Moon landings and you will soon see images of Mission Control – a huge room packed with consoles, giant screens and people wearing headsets. This impressive set-up lay at the heart of the space programme. The mission controllers never donned a spacesuit or travelled to a distant world. Instead, they wore everyday clothes, kept their feet firmly on the ground and often hid away from the public gaze. Nevertheless, they were central to the success of the entire enterprise.

Imagine that you are living in the early 1960s and have been charged with putting a person on the Moon by the end of the decade. The whole world will be watching and the nation's reputation is on the line. What sort of people would you recruit for Mission Control? Maybe you would go for highly experienced scientists and engineers. Or people who have graduated from the country's most prestigious universities and colleges. Amazingly, hardly anyone in Mission Control had any of these attributes. Instead, most of the controllers came from modest, working-class backgrounds, and were often the first in their families to go to college. Perhaps most surprising of all, they were astonishingly young. In fact, when Neil Armstrong set foot on the Moon, the average age of the mission controllers was just twenty-six years. The group had few of the qualities that we commonly associate with success and yet, somehow, they managed to accomplish the seemingly impossible.

I wanted to discover why, from a psychological perspective, Mission Control was such a hotbed of success, and I was fortunate enough to be given the opportunity to interview several key controllers. Now in their seventies and eighties, they were all remarkably generous with their time and thoughts. My interviewees had seen history in the making and had fascinating stories to tell. They were also a great deal of fun. During one interview, for instance, I asked a controller whether it was true that the group was trained to communicate using the fewest number of words possible. After a brief pause, he replied, 'Yes'.

I eventually discovered that their breathtaking achievement was due to a unique mindset. Combining the interviews with mission archives and academic research, I identified the eight principles that I believe make up this remarkable and highly effective approach to life. You are about to discover how Kennedy's dream became a reality. Along the way, you will relive historic events and rub shoulders with a group of ordinary people who achieved the extraordinary. During the journey you will encounter the key psychological principles at play, including how the seeds of success were sown in the President's charismatic speeches, how pessimism was crucial to progress, and how fear and tragedy were transformed into hope and optimism.

Perhaps most important of all, you will discover a series of practical techniques that you can use to incorporate these winning principles into both your professional and personal life. Whether you want to start a business venture, change careers, find your perfect partner, raise a loving family, get promoted, gain a new qualification, escape the rat race or pursue a lifelong passion, these techniques will help you to reach your own Moon.

1
WE CHOOSE TO GO TO THE MOON . . .

Where we discover how an entire nation fell in love
with a mission to the Moon, and find out how you
can harness the power of passion.

In October 1957, America's CBS television broadcast the first episode of what was to become an iconic sitcom, *Leave it to Beaver*. The programme celebrated 1950s American family life and revolved around the adventures of eight-year-old Theodore 'Beaver' Cleaver.

Getting the programme on air hadn't been easy. The producers had initially wanted to open with an episode in which Beaver ordered a pet alligator, and then hid his new reptilian pal in the toilet tank of his parents' bathroom. However, broadcasting guidelines recommended that bathrooms and toilets weren't shown on television, and the CBS chiefs became nervous about airing the episode. After much wrangling, the producers offered to re-edit the alligator episode and only to include a few shots of the offending toilet. Unfortunately, the editing took longer than expected and so the network was forced to air an alternative debut episode.

Oblivious of the controversy, millions of Americans came home from work on Friday 4 October 1957 and chuckled away as Beaver became convinced that he was going to be expelled from school, hid up a tree and then attempted to make amends by presenting his teacher with a rubber shrunken head. It all seemed like harmless fun. In reality, the programme's safe and secure worldview was just days away from being blown apart for ever.

A few hours before *Leave it to Beaver* had aired, some Americans had noticed a tiny ball of light moving rapidly through the sky. Around the same time, amateur radio enthusiasts had started to pick up a series of deeply strange beeps. Word of these odd happenings quickly spread and within days the public started to worry. Many believed that the peculiar sights and sounds were due to a new type of shooting star. Others thought that an alien invasion was imminent. Some put it all down to hallucination and hysteria. In fact, the truth was far more troubling.

Earlier that day, the Soviets had placed the first man-made object into orbit around the Earth. Sputnik was about the size of a basketball, weighed around eighty kilograms and travelled at an astonishing 29,000 kilometres per hour. It flew a few hundred kilometres above the Earth's surface, took about ninety minutes to orbit the planet and passed over America several times each day. The Soviets were eager to ensure that their high-tech space ball spooked America. Spherical and highly polished, they had designed Sputnik to reflect as much sunlight as possible and so be visible from the ground. Eager to extend the sense of mystery, the Soviets didn't release photographs of their silver satellite until several days after the launch.

The Soviet plan worked perfectly. America panicked, and questions came thick and fast. How had a totalitarian regime managed to outfox the world's mightiest democracy? Were Uncle Sam's secrets safe from this new eye in the sky? Was the rocket used to launch Sputnik capable of carrying a nuclear warhead? The Cold War had reached new heights and the space race had begun.

Seen from a psychological perspective, Sputnik is fascinating. Speak to anyone who lived through the crisis and they will quickly talk about the genuine sense of fear created by the little beeping ball. Suddenly their future seemed unpredictable and uncertain. Desperate, they turned to the President in the hope of strong leadership and a steady hand on the rudder. Unfortunately, it was not to be.

HOW NOT TO RESPOND TO A CRISIS

President Eisenhower had left Washington, D.C. for a golfing weekend on the morning of the Sputnik launch and didn't hold a press conference about the mysterious orb until five days later.[3] When he eventually appeared before journalists, the President played down the threat and appealed for calm. Fellow politicians became worried that their leader was out of touch with reality, with one senator pleading with the President to step up to the mark and declare a national 'week of shame and danger'. The Governor of Michigan, Gerhard Mennen Williams, even published a sarcastic poem in the *New York Times* criticizing Eisenhower:[4]

7

'Sputnik'
Oh little Sputnik, flying high
With made-in-Moscow beep,
You tell the world it's a Commie sky
and Uncle Sam's asleep.

You say on fairway and on rough
The Kremlin knows it all,
We hope our golfer knows enough
To get us on the ball.

A few days later, the second episode of *Leave it to Beaver*, featuring the pet alligator, aired, and the sitcom became the first national television programme to show images of a toilet on the small screen. Unfortunately, this milestone in broadcasting history was overshadowed by Sputnik. By then, the little beguiling beeping ball was starting to make its mark on almost every aspect of society, with bartenders creating Sputnik cocktails and toy manufacturers flooding stores with model satellites and space suits.[5]

Three weeks after the satellite's launch, its batteries finally ran out and the beeping ceased. Nevertheless, America's sense of concern continued to grow. Some politicians argued that the country's preoccupation with the good life had caused people to take their eye off of the (silver) ball and to value home comforts over national security. As Republican Senator Styles Bridges forcibly put it:

'The time has clearly come to be less concerned with the depth of the pile on the new broadloom rug, or the height of the tail-fin on a new car, and to be more prepared to shed blood, sweat, and tears if this country and the free world are to survive.'[6]

On the other side of the globe, a very different story was fast emerging. The Soviet leadership was delighted to see the way in which Sputnik had spooked America. Eager to follow up on their success, the government quickly ensured that postage stamps, posters and magazine covers all carried images of Sputnik. In addition, the nation's top rocket scientists were quickly given the green light for the next stage of their space programme.

A month later, the situation for America went from bad to worse. To help celebrate the fortieth anniversary of the Bolshevik Revolution, the Soviets successfully sent the first living creature into orbit around the Earth. Sputnik 2 weighed about five times as much as its predecessor and carried a small dog named Laika. Soviet scientists had originally planned to build a spacecraft that could return Laika safely to Earth, but had to launch a one-way mission to meet the anniversary deadline. Unfortunately, Laika died from overheating a few hours after the launch, but the flight demonstrated that animals could survive weightlessness and suggested that the Soviets might soon send a human into space. Once again, the West was stunned.

On 6 December 1957, millions of Americans tuned into the tenth episode of *Leave it to Beaver* and watched the show's diminutive hero publicly humiliated at a school dance. Earlier that day, the entire country had found itself in a conceptually similar situation. Under the watchful eye of the world media, America had attempted to launch its own version of the Sputnik satellite. Their Vanguard TV-3 rocket slowly lifted off from the launch pad, then rapidly returned to Earth and vanished in a colossal fireball. Newspaper editors across the globe had a field day – the *Daily Express*: 'Kaputnik'; the *Daily Herald*: 'Flopnik'; *News Chronicle*:

'Stayputnik'. Tongue in cheek, the Soviets suggested that America might be eligible for United Nations aid earmarked for undeveloped countries.[7]

Something had to be done, and quickly.

DARING TO BE BIG, BOLD AND FIRST

In 1958, the Eisenhower administration established the National Aeronautics and Space Administration (NASA), ploughing millions of dollars into science education. Two years later, John F. Kennedy went head to head with Richard Nixon for the American Presidency. The space race took centre stage, with Kennedy pledging that he would do his best to ensure that America would be first across the finishing line.[8] Kennedy won the day, beating Nixon to the presidency with one of the smallest margins in American history.

Desperate to win over a divided nation, the young President wanted to deliver an impressive inauguration address; he turned to his top speechwriter for help. Ted Sorensen was a highly talented wordsmith with a gift for capturing big thoughts in small phrases. On 29 January 1961, 44-year-old Kennedy became one of the youngest ever Presidents of the United States of America. It was a bitterly cold day and a blanket of snow had brought the nation's capital to a virtual standstill. Nevertheless, hundreds of thousands of people watched as Kennedy stepped up to the podium and delivered what is widely considered to be one of the most impressive political speeches in history.

Kennedy had energy and determination and spoke to a post-war generation that seemed to have lost its way. The young President encouraged every American to reflect on how they could help others and emphasized the value of public service. His fourteen-minute inaugural speech ended with a now famous phrase that beautifully encapsulated Kennedy's bold vision: 'Ask not what your country can do for you, ask what you can do for your country.'

Despite Kennedy's sense of new-found optimism, his presidency did not start well. In early 1961, Uncle Sam faced further humiliation when both the Cuban Bay of Pigs invasion and several more rockets failed to get off the ground. Soon after settling into the White House, Kennedy had turned his attention to the space race by starting to review the possible plans that had been prepared by the country's top scientists and engineers.[9] The President knew that he needed a vision that would engage the hearts and minds of millions of people. Some of the experts suggested that America go head to head with the Soviets and attempt to launch a huge satellite. Others preferred the idea of building a giant space station that would permanently orbit the Earth. Kennedy's preference was to think bigger and bolder still.

After many months of meetings, the President eventually supported an idea that resonated with his grand sensibilities: the world's first manned mission to the Moon. Mindful of the Russians beating America to the punch, Kennedy added a tough time frame, declaring that he wanted a man to walk on the lunar surface before the end of the decade. It was an astonishingly audacious aim.

For Kennedy, this ambitious idea ticked all the boxes. Going to the Moon would be a remarkable first and also would leave a lasting mark on history. Moreover, it would bloody the nose of the Soviets and help America to pull ahead in the space race. Perhaps most important of all, the President believed that it would ensure that a democratic power controlled the heavens and so would make the world a more peaceful place. There was, however, just one significant problem. Getting to the Moon was going to cost a fortune and Kennedy had to first convince Congress that this was a prize worth paying for.

In May 1961, Kennedy appeared before a joint session of Congress to deliver a special message on 'urgent national needs'. The President outlined his spectacular vision to place a man on the Moon within the decade, and calmly explained that the programme would involve immense technological challenges and demand eye-watering levels of public funding. However, Kennedy was firm in his resolve, noting that anything less than a man on the Moon wouldn't cut the mustard:

'I am asking the Congress and the country to accept a firm commitment to a new course of action . . . If we are to go only half way, or reduce our sights in the face of difficulty, in my judgment it would be better not to go at all.'

The President then declared that all of America was going to have to work together if this audacious goal were to stand any chance of success. For Kennedy, this was not about sending one person to the Moon, but rather an entire nation reaching for the stars.

Kennedy ended his speech by asking Congress to dig deep and make available the $7 to $9 billion needed to fund the work. (In reality, the programme would eventually cost an estimated $25 billion and, at its height, consume over 5% of annual public expenditure of the entire country).

The President was concerned that Congress might reject his

proposal or be tempted to commit to a much smaller spend, but his brave and bold vision won the day and the motion was carried after just an hour of debate.

He had won the support of Congress, but would he be able to energize and engage the American public?

'WE CHOOSE TO GO TO THE MOON': THE POWER OF PASSION

Throughout 1961, a space-based taskforce searched for a suitable site for mission headquarters, and eventually decided to build their Manned Spacecraft Center on land close to Rice University in Houston, Texas. On 12 September 1962, Kennedy travelled to Rice University's football stadium to announce his dream of humans reaching the Moon before the decade was out.

The event attracted an audience of over 40,000 people. Sitting up in the stands that day was a fifteen-year-old schoolboy named Terry O'Rourke. Now in his seventies, Terry still remembers seeing Kennedy: 'You live thousands of days in your life, but I still have a vivid memory of those few hours. I took time off school and rode my bicycle over to Rice Stadium. Back then there was very little security, and so I just went in and sat down. Boy it was a hot day. I can remember everyone struggling in the really humid, semi-tropical, heat.'[10]

Again, Kennedy had worked with speechwriter Ted Sorensen to craft an address that they hoped would enchant and entice the nation. Looking down from the stand, Terry saw the President step up to the podium and deliver his opening words:

'We meet at a college noted for knowledge, in a city noted for progress, in a State noted for strength, and we stand in need of all three.'

Terry immediately felt the power of Kennedy's words and presence: 'You have to remember that this was during the Cold War, and everyone felt apprehensive and scared. We had no idea how the Soviets could have been ahead of us in the space race. And here was Kennedy – handsome, smart, and charismatic – telling us that there was still hope.'

Early in his speech, Kennedy outlined the vision that had galvanized Congress. America would put a man on the Moon before the end of the decade. Next, Kennedy focused on the thrill of being a trail-blazing pioneer and the importance of the space race for the future of humanity:

Those who came before us made certain that this country rode the first waves of the industrial revolution, the first waves of modern invention, and the first wave of nuclear power, and this generation does not intend to founder in the backwash of the coming age of space. We mean to be a part of it – we mean to lead it. For the eyes of the world now look into space, to the Moon and to the planets beyond, and we have vowed that we shall not see it governed by a hostile flag of conquest, but by a banner of freedom and peace. We have vowed that we shall not see space filled with weapons of mass destruction, but with instruments of knowledge and understanding.

Rice University was a small institution compared to the nearby

University of Texas at Austin. For years, there had been a fierce rivalry between the Universities' football teams, with The Texas Longhorns regularly beating the Rice Owls. Kennedy received the biggest cheer of the day by improvising a line about this rivalry, reflecting on the importance of embracing difficult challenges:

But why, some say, the Moon? Why choose this as our goal? And they may well ask why climb the highest mountain? Why, 35 years ago, fly the Atlantic? Why does Rice play Texas?

We choose to go to the Moon in this decade and do the other things, not because they are easy, but because they are hard, because that goal will serve to organize and measure the best of our energies and skills, because that challenge is one that we are willing to accept, one we are unwilling to postpone, and one which we intend to win.

Terry remembers the stadium being energized by Kennedy's sense of passion and enthusiasm: 'I had 100% confidence in him. He said we were going to go to the Moon and I totally believed him. We all believed him. You can call it optimism, or arrogance, or innocence, but it seemed as if everyone in that stadium really thought that America could pull it off.'

Kennedy then turned his attention away from the 'why' of putting a man on the Moon and focused on the 'how'. He admitted that America was currently behind in terms of space technology and reflected on the immense technological barriers that would have to be overcome. Getting to the Moon, declared the President, would involve constructing a giant rocket the length of the stadium's football field, designing instrumentation that would outperform the world's finest watches and developing materials capable of withstanding heat about half that of the temperature of the sun (with Kennedy improvising another line: 'almost as hot as it is here today').

Kennedy ended his historic address by comparing the journey ahead with the expeditions made by one of the world's great explorers:

The great British explorer George Mallory, who was to die on Mount Everest, was asked why did he want to climb it. He said, 'Because it is there.' Well, space is there, and we're going to climb it, and the Moon and the planets are there, and new hopes for knowledge and peace are there. And, therefore, as we set sail, we ask God's blessing on the most hazardous and dangerous and greatest adventure on which man has ever embarked.

Kennedy's comments about the importance of making the world a better place resonated with the then fifteen-year-old Terry O'Rourke: 'His speech really touched me. The whole idea of doing something for the greater good was completely fused in my mind that day, and I came away wanting to help my nation and fellow humans in the best way possible.'

Soon after leaving Rice Stadium, Terry wrote to his local Congressmen and was invited to become a pageboy in the House of Representatives. A year later, he experienced another unforgettable moment, coming face to face with Kennedy on the lawn of the White House. Terry went on to study law, and has enjoyed a long and distinguished legal career, dedicating his life to fighting social injustice and to environmental protection. In line with Kennedy's goal of working for the greater good, Terry was a part of the presidential team that persuaded Congress to create the US Department of Energy; he also served on the White House Staff of President Jimmy Carter.

Terry wasn't the only one in the stadium to have his life changed by Kennedy's speech. Standing a short distance away from Terry was Rice University student and ace basketball player Jerry Woodfill.[11]

Originally from Indiana, Jerry had developed a childhood fascination with basketball and eventually won an athletics scholarship to Rice University. Jerry went along to hear Kennedy speak and, like Terry O'Rourke, can remember the blistering heat. At the time, Jerry's life was not going especially well. He hadn't obtained great marks in his exams (two Cs, two Ds, and an F minus) and his basketball training was proving tough. As Kennedy started to speak, Jerry felt something inside him stir. By the end of the speech, he was a changed man. Energized by Kennedy's passionate vision of being the first to go to the Moon, Jerry returned to Rice University and dropped his basketball career, instead devoting himself to studying electrical engineering. When he graduated, Jerry applied to NASA and was invited to help develop the safety systems on the spacecraft that would attempt to land on the Moon. On 20 July 1969, just seven years after seeing Kennedy speak at Rice University, Jerry was working in the Manned Spacecraft Center, and helping Neil Armstrong and Buzz Aldrin set foot on the lunar surface.

Millions of people across America were inspired and enthused by Kennedy's vision of going to the Moon, and soon it seemed as if the whole nation was suffering from space fever. In just a few months, the President had managed to motivate politicians, the public, scientists, and engineers alike. America had found its dream destination and humanity was on its way to the Moon.

*

Harnessing the Power of Passion

The idea of going to the Moon had energized millions of people around the world. Some of them had spent their childhoods reading about the adventures of Flash Gordon and Buck Rogers and loved the idea of exploring space. Others wanted to head to the Moon because it was such a difficult, novel and daring challenge. Then there were those that were driven by a strong sense of purpose. Like Kennedy, they believed that heading into the heavens would promote freedom and democracy, and so help create a better world for future generations. Finally, some enjoyed the sense of competition created by the space race and were eager to cross the finishing line ahead of the Soviets.

The same sense of passion motivated many of the scientists and engineers who would eventually transform Kennedy's vision into a reality, and it was essential to their eventual success. It transformed work into play and helped those working to cope with the long hours and tough deadlines. Bill Tindall Jr. was one of the most senior engineers involved in the Apollo programme. Several years after the Moon landings, Tindall was asked what had motivated so many people to work so hard to make it all happen. Tindall objected to the use of the word 'work' and noted:

'I would just change the word from work to play because I never thought we were working at all. And that is the honest to God truth. It was just so much fun.'[12]

The same applies to many of those in Mission Control. When Flight Director Glynn Lunney was asked about how he felt about being part of the team that eventually put Neil Armstrong on the Moon, he responded: 'We loved it. We loved the work, we loved the comradeship, we loved the competition, we loved the sense of doing something that was important to our fellow Americans.'[13] Similarly, Flight Controller Steve Bales said that being part of the Apollo missions was so exciting and so much fun, that he would have worked on the programme even if he had only been paid enough to cover his living expenses.[14] And when Mission Control Flight Director Gerry Griffin was asked whether he would be prepared to face the high workloads and incredible stress again, he instantly replied: 'Of course. I am just sad that it ever had to end.'[15]

The controllers' comments are supported by a wealth of scientific evidence. Robert Vallerand, from the University of Quebec, has produced hundreds of academic papers on the psychology of passion.[16] After studying the lives and minds of thousands of passionate people, Vallerand has discovered that this often overlooked factor is one of the main secrets of success. When people do what they love, their work feels more like play and they are more likely to keep going when the going gets tough. As a result, they end up being especially productive and successful. In the same way that Kennedy energized an entire nation by announcing his intent to go to the Moon, passion has the power to propel people to incredible heights in both their personal and professional lives.

When it comes to finding your dream destination, follow your passion. Or if you are forced to head along a certain path, find a way to become more passionate about the journey. Unfortunately, a surprisingly large number of people struggle to identify exactly what makes their eyes sparkle and their life worth living. The following techniques are designed to inject some more passion into your life, so providing the fuel for getting to your own Moon. These techniques are based around the factors that made people passionate about Kennedy's dream; they involve you answering nine vitally important questions, understanding the science of supersizing, making today the most important day of your life and creating your very own space race.

NINE QUESTIONS

Many of the scientists and engineers working on the Moon landings had a lifelong passion for flight and space exploration. Similarly, most people are naturally passionate about something. From painting to pottery, music to mosaics and cloud-watching to conjuring, this fascination often develops when they are young and has the potential to float their boat throughout the rest of their lives (especially if they are passionate about sailing). Unfortunately, when life gets complicated and busy, people often forget what puts a spring in their step. If that sounds like you, use this technique to identify your natural passions.

First, find a quiet spot and jot down the answer to the following nine questions.

1) List three moments in your life when you felt especially excited, enthusiastic and alive.
2) Picture yourself locked in a room where you are only allowed to read books and magazines about one topic. Which topic would you choose?

3) Imagine that you are financially secure and therefore free to do whatever you like with your life. After travelling the world, buying a house or two, supporting deserving friends and family and donating to your favourite charities, what would you do with your life?

4) What did you love doing when you were a child? Are there any childhood toys or objects that you have held on to over the years? If so, why?

5) What hobbies and interests did you once enjoy but are now not part of your life?

6) Pretend to yourself that that you are in your twilight years. Look back over your life and think about how you wished you had spent the last thirty years. What regrets do you have? What do you wish you had done?

7) Imagine that you can create something new. It can be anything at all. Maybe it's a new type of wheelbarrow, a new superhero, a new website or a new way of learning to play the guitar. What would you create?

8) Have you ever been engaged in an activity and suddenly noticed that time has whizzed by? Maybe you thought that you had been working away for thirty minutes, only to discover that a few hours have passed. What were you doing at the time?

9) Imagine being given a large board and asked to cover it in pictures that appeal to you. You are allowed to stick any photograph, drawing or image that you like on the board. What kind of pictures would you place on it?

As you might have guessed, these nine questions are designed to help you find the love of your life. In a moment, you will be asked to review your answers and discover your passion by identifying the general themes that have emerged. However, just before you start, please think about the following two points.

- **Don't worry about maxing out on passion.** Psychologist Benjamin Schellenberg asked over a thousand students to indicate their happiness, physical wellbeing and cheerfulness, along with whether they had none, one or two passions in their lives.[17] People with two passions were the happiest of all. Worried that these people might be devoting more time to activities they enjoy, the researchers also asked everyone to indicate how much time they spent indulging in their passions. People with two passions were happier, even when the time they invested in their two passions was the same as the time invested by those with one passion. The message is clear – having one passion is good, but having two or more is even better.
- **A quick note of caution – don't overdo it.** Research shows that not all passion is positive. Some people run the risk of becoming obsessed with a passion, such that it starts to control their lives. Often they will feel as if they cannot stop doing whatever they are obsessing over, and are sometimes driven on by external rewards, such as praise, fame or money, rather than enjoying the activity for its own sake. This type of obsession can lead to burnout and even injury with, for instance, dancers performing even though they are injured or cyclists going out in dangerous weather conditions. Make sure that you're passionate – not obsessive.

OK, take a look at your answers to the nine questions and try to identify what you really care about in life and what puts a spring in your step. Go.

Perhaps, for instance, you frequently mentioned tap-dancing, learning as much as possible about the Battle of Hastings, painting, creating smartphone apps or making pottery, going metal detecting or visiting the theatre. Either way, now think about identifying your dream destination by figuring out how can you use your passion to create a new ambition or goal. Once again, before you start, consider the following two points.

First, many people want to make a living by doing what they love. If you are struggling to think of how that might happen, think about how you might be able to incorporate your passion into your existing work. For instance, if you work in recruitment, but have a passion for technology, you might suggest heading up a working group to explore the potential of social media to attract new personnel. Or if you work in customer service, but have a passion for acting, can you use your performing skills to help build a rapport with customers?

Second, if you do have several passions, think about how you might combine them to create a unique offering. For instance, if you have a green thumb and are into mathematics, could you become a landscape gardener specializing in geometrical patterns? Or if you are into making music and fitness, could you create tunes that motivate people to work out at the gym? Or perhaps you could follow in the footsteps of Apollo astronaut Al Bean. Bean both walked on the Moon and is a skilled artist; he now paints pictures of the lunar surface incorporating actual Moon dust.

OK, now try to find your dream destination.

The Science of Supersizing

*'Make no small plans. They have no power to stir
men's blood. Go big or go home.'*

– Daniel Burnham

When Kennedy announced that America would attempt to go to the Moon, he set the bar extraordinarily high. This type of audacious aim – often referred to as a 'stretch goal' – can sow the seeds of success by making people more passionate, disrupting complacency, promoting innovation, elevating aspirations, and broadening horizons. The approach has proved to be highly effective in several well-known organizations. For instance, Steve Jobs helped Apple reach new heights by using his 'reality distortion field' – a mix of charm and vision that convinced people they could achieve the seemingly impossible. Similarly, the highly successful inventor and entrepreneur Elon Musk frequently creates amazing technological leaps by using 'overly optimistic deadlines'.

Research suggests that we can all benefit from the passion that flows from having goals that are big, audacious and seemingly impossible. Rather than trying to start a small business, for instance, it's better to imagine creating an empire. Instead of trying to have a positive impact on a small community, it's more effective to aim to help millions. And instead of settling for being able to run a short distance, it's better to set your sights on completing a marathon. To make any stretch goal especially effective, follow these simple rules:

- First, stretch goals work best when they evoke the 'fear factor', producing a rush of adrenaline that scares, excites and energizes in equal measure. This rush has its basis in an emotional conflict. The best stretch goals usually have no obvious way of being achieved at first and so are associated with uncertainty and fear. However, at the same time they feel expansive and necessary, which creates a positive feeling of optimism and hope.

- Second, when it comes to stretching, there's a sweet spot. If you create a goal that is ambitious, but relatively achievable, you are not really challenging yourself or your organization. Worse still, when you are successful you run the risk of regret when you realize that you could have achieved so much more. At the opposite end of the spectrum, an utterly unrealistic goal ('I will become the next President of the United States of America within a year') can become a source of discouragement and, perhaps more importantly, cause you to set your sights too low next time around. To prevent this, assess your chances of realistically achieving the goal if you are fully committed. Before speaking at Houston, for instance, President Kennedy consulted rocket scientists and space engineers about the possibility of putting a man on the Moon before the end of the decade. They told him that doing so was a long shot (quite literally), but possible. If you are 90% certain that a goal can be achieved, then it's almost certainly too easy. Equally, a 10% chance is probably setting the bar too high. Some researchers believe that it is best to aim for around a 50% to 70% chance of success.

What's your audacious and ambitious aim? Think big and be first. Go further or faster than anyone else. Use the fear factor, and the '50:70%' rule, to create your perfect goal. After setting a stretch goal, some organizations and individuals succeed and others don't. Either way, the audacious vision propels them on their way, and ensures that even those who do not fully succeed are far more successful than those setting the bar low. Les Brown once famously observed: 'Shoot for the Moon. Even if you miss, you will land among the stars.'

Moonshot Memo
Subject: SMarT thinking

People are more likely to achieve their goals when they use SMarT thinking – making their aims Specific, Measurable and Time-constrained.[18] Kennedy's vision was extremely SMarT. He didn't say that America was simply going to reach for the stars or head into space. Instead, he promised that the nation would land on the Moon and return safely to the Earth (both specific and measurable) by the end of the decade (a time-constrained endpoint).

Increase your chances of success by using some SMarT thinking, and making your aims and ambitions Specific, Measurable and Time-constrained. Oh, and make a note of your goals. People who keep a journal of their aims and ambitions are about 30% more likely to make them a reality.

Finally, keep it simple. One Apollo astronaut remarked that Kennedy's entire vision could be summed up in just three words: 'Man. Moon. 1970.' Can you sum up your goal in just a handful of words?

Make Today the Most Important Day of Your Life

'The two most important days in your life are the day
you are born, and the day you find out why.'[19]
– Ernest T. Campbell

Despite having relatively low wages, scrubbing enclosures and scooping up animal waste, psychologists have discovered that zoo-keepers are some of the most content employees in the world.[20] Why? Because they can see how their hard work makes the world a better place and so are more prepared to put in the long hours and endure the occasionally terrible smell. In short, they enjoy their working lives because they have a sense of porpoise!

Similarly, many of those working on the Apollo programme were driven by a sense of purpose: they believed that getting to the Moon would help create a better world by promoting freedom and democracy.

Research conducted by University of Pennsylvania psychologist Adam Grant shows that even a small sprinkling of purpose can make a surprisingly big difference.[21] Grant's University ran a call centre. Day after day, the staff rang former graduates and asked them if they would like to contribute to a scholarship fund for future students. It was highly repetitive work and the staff frequently faced rejection. In an attempt to add meaning to the mundane, Grant tracked down a student who had benefited from a scholarship and asked the student to spend a few minutes telling the staff how the grant had transformed their life. This simple change worked wonders. The staff suddenly understood why they were doing what they were doing and so found their job more meaningful. Grant tracked the staff's performance over the next few weeks and was astonished to see them spending 142% more time on the telephone, and raising 171% more revenue.

How can you inject more meaning into your life? Perhaps the easiest route involves finding ways of helping others, serving a greater good or making the world a better place. Remember how Terry O'Rourke heard Kennedy's speech about the importance of creating a better future and went on to dedicate his life to promoting social justice and environmental protection? As the writer Hunter S. Thompson once remarked: 'Anything that gets your blood racing is probably worth doing.' What gets your blood pumping? What problem in the world would you most like to fix?

The same idea can also help you to instantly add more meaning to your work. Psychologist Amy Wrzesniewski is Professor of Organizational Behavior at the Yale School of Management; she has devoted much of her career to helping people find meaning in almost any occupation.[22] Her approach is referred to as 'job crafting' and involves several simple techniques that help transform the job you have into the job you love.

Perhaps the easiest approach involves you asking yourself one simple question: 'How does my job help others?' This is possible in practically any occupation and just takes a little effort to move beyond the usual job description. For instance, teachers might focus on how their community benefits from

28

well-educated children, or mobile-phone designers might think about how their products help people bond together and share happy memories; similarly, supermarket checkout assistants might remind themselves that they provide lonely customers with a brief, but important, moment of social contact.

According to one story, Kennedy once visited the Manned Spacecraft Center and asked a cleaner there to describe his job. The cleaner replied: 'I am helping to put a man on the Moon.' Apocryphal or not, the story captures the genuine ethos of those reaching for the Moon. The thousands of people involved in the Apollo programme weren't only designing a rocket engine, tightening a bolt or cleaning a floor. Instead, they saw themselves as making a vital contribution to an important enterprise. Adopt the same attitude and encourage it in those around you, and you will soon discover how passion flows from purpose.

How can you add more meaning to your life? How can you contribute to a greater good, and make the world a better place? What makes your blood race or your heart sing, and what can you do about it? And remember, you can inject an instant sense of purpose into any activity, by asking one simple question: 'How does this help others?'

CREATE YOUR OWN SPACE RACE

In 1898, one of the founding fathers of modern-day psychology discovered something strange. By day, Professor Norman Triplett was a well-respected psychologist at Indiana University. However, the rest of the time he was a passionate advocate of the then latest fad of cycling. Triplett decided to combine his love of the human mind with his joy of cycling and conducted a groundbreaking study that kick-started the whole of sports psychology.[23]

Triplett examined race times from the records of the Racing Board of the League of American Wheelmen and discovered something odd. When racing against their fellow Wheelmen, cyclists achieved much faster times than when they were trying to beat the clock on their own. Over the last hundred years or so, this effect has been replicated many times, with research showing that having people compete against others for a prize, no matter how trivial, boosts performance.[24] The effect is especially powerful when people work together in a team and when they are aware of the ongoing performance of both themselves and their competitors.

Some of the most recent work in the area has shown that rivalry is especially important. Professor Gavin Kilduff, from New York University, was curious to discover whether people are especially competitive when they compete against someone known to them and whom they consider a rival.[25] Following in the tyre tracks of Triplett, Kilduff analysed archival data from over a hundred long-distance running races, and discovered that the presence of a rival caused runners to shave off vital seconds during a race. Other work even suggests that merely imagining that you are competing against a rival can help boost performance.

The Americans weren't just trying to get to the Moon. Instead, they were trying to beat their Russian rivals, and the resulting sense of competition energized them and made them especially motivated to win.

The next time you want a quick boost of passion, consider creating your own space race. Who are your competitors and main rivals? Who would you like to beat? Can you motivate yourself and others by creating a fun contest? For instance, could you create a competition between different parts of an organization to see who can recycle the most? Or compete with your partner to see who can lose the most weight? Or motivate yourself at the gym by imagining yourself competing against your rival? Either way, boost your passions by getting in touch with your inner competitive spirit.

SUMMARY

Follow your passion. Alternatively, if you have to head in a certain direction, discover how to become more passionate about the journey.

– To uncover your hidden passion, think about what you used to do as a child, your hobbies and interests, which books and magazines you would take to your desert island and which activities would turn hours into minutes.

– Think big and be first. Kennedy's goal set the world alight because it was so audacious and ambitious. What is your bold, brave and exciting goal? How will you be first?

– Make life more meaningful by thinking about contributing to the greater good. What makes your blood race and what can you do about it? To inject a sense of purpose into any activity, ask yourself one simple question: 'How does this help others?'

– And finally, give yourself a quick boost of passion by creating your own space race. Find a way of turning an activity into a fun competition or game, and by creating a light-hearted sense of rivalry.

2

'JOHN, IT WORKED BEAUTIFULLY'

Where we meet the innovative engineer who saved the day with a brilliant mission plan, and find out how you can create ideas that are out of this world.

Kennedy had made an entire nation passionate about going to the Moon before the decade was out. The President had, however, left out one important detail – how on earth this audacious goal was to become reality in such a short time frame. Strangely, almost one hundred years before, a French novelist had tackled exactly this issue. A century on, his pioneering work shaped the thoughts of modern-day rocket scientists.

Jules Verne was born in 1828 on a small artificial island in Nantes. He was passionate about writing from a young age. Verne wanted his stories to be as scientifically accurate as possible, and spent vast amounts of time reading about the latest technological breakthroughs at the National Library of France. Eventually developing a love of fantastical travel writing, he produced several well-known novels, including *Journey to the Centre of the Earth,*

Twenty Thousand Leagues Under The Sea and *Around The World In Eighty Days*. In 1865, Verne turned his attention towards space and produced a groundbreaking, light-hearted, science-fiction novel entitled *From the Earth to the Moon*.

Verne's story is set just after the end of the American Civil War and centres on the fortunes of the Gun Club in Baltimore. The Club is dedicated to the design of weaponry, with most of its members having served on the battlefield. At the start of the story, Verne paints a vivid and humorous picture of the injuries sustained by this motley and eccentric crew:

Crutches, wooden legs, artificial arms, steel hooks, caoutchouc jaws, silver craniums, platinum noses, were all to be found in the collection; and it was calculated by the great statistician Pitcairn that throughout the Gun Club there was not quite one arm between four persons and two legs between six.

The end of the Civil War has reduced the nation's interest in weapons and the President of the Gun Club announces that the Club should reverse its ailing fortunes by building a giant cannon and firing a capsule at the Moon. A French adventurer offers to travel on board the capsule, and manages to persuade the Club President and an army Captain to join him. The Club's ambitious scheme is reported in newspapers and magazines across the globe, resulting in a flood of financial donations. Eventually, the Gun Club raises an impressive $5.5 million, with Americans donating $4 million and the English failing to give 'a single farthing'.

The Gun Club name their giant cannon the Columbiad, deciding to construct it by digging a huge hole in the ground and lining the hole with cast iron. After carefully considering several potential sites across America, the members eventually settle on a location just south of Tampa, Florida. All goes well with the construction and soon the 900-foot long by 9-foot wide cannon is ready for action.

Realizing that their capsule needs to weigh as little as possible, the members decide to build it from aluminium, with one Gun Club member remarking that this recently discovered metal 'seems to have been created for the express purpose of furnishing us with the material for our projectile'. After carefully calculating the force required to escape the Earth's gravitational pull, 400,000 pounds of guncotton are stuffed into the base of the cannon.

On launch day, the three adventurers bravely descend down the barrel of the huge cannon and enter their bullet-shaped aluminium capsule. As five million bystanders merrily sing 'Yankee Doodle', the capsule is blasted successfully into space and the world's first astronauts begin their journey to the Moon.

Verne's story ends on a dramatic cliffhanger. Astronomers track the capsule's journey through a giant telescope, watching in horror as the three men miss the Moon and become trapped in lunar orbit. In the final paragraph, one of the astronomers gloomily remarks that the brave adventurers are likely to end their days circling the Moon, while a member of the Gun Club strikes a more optimistic note: 'Those three men have carried into space all the resources of art, science and industry. With that, one can do anything; and you will see that, some day, they will come out all right.'

From the Earth to the Moon proved a huge success and Verne returned to the saga in his sequel *Around the Moon*. At the start of this second story, readers discover that the capsule actually contained the three astronauts, two dogs (Diana and Satellite), half a dozen chickens and a fine cock. Shortly after the dramatic launch, the men enjoy some wine and soup, but then discover that Satellite was hurt when the capsule was blasted out of the giant cannon. Unfortunately, the dog's injuries prove fatal and the three men throw Satellite out of the capsule's window (thus running the risk of turning Satellite into a genuine satellite). The gravitational force from a passing asteroid then throws the capsule off course and causes the adventurers to enter lunar orbit. After surveying the surface of the Moon with a pair of opera glasses, the capsule encounters another course-altering asteroid and zooms back towards the Earth. The three men eventually splash down in the Pacific Ocean, are rescued by the American Navy and enjoy a lavish homecoming.

Before writing his novels, Verne consulted his cousin, who was a professor of mathematics, asking him to help make the stories as scientifically sound as possible. Both of these Verne novels are filled with complex equations and formulae and contain entire chapters dedicated to rocketry, vacuums and weightlessness. As a result, the novels represent the first comprehensive attempt to figure out the mathematics and physics of space flight and astronautics.

Many of Verne's projections and predictions were astonishingly accurate. He was, for instance, the first person to present the correct calculation of the acceleration required to escape the Earth's gravitational pull ('12,000 yards per second'), and specifies the capsule's trajectory so accurately that space historians were later able to map out its exact path to the Moon. In addition, Verne presents pioneering perspectives on the possible effects of weightlessness (although he wrongly thought that the crew would only experience it halfway through their journey to the Moon), and foresaw the use of 'retrorockets' (engines that provide thrust in the opposite direction to the motion of the spacecraft, thus causing it to decelerate).

Moonshot Memo
Subject: Verne

Strangely, many aspects of Verne's novels closely mirror the Apollo missions . . .

– In his first story, a team of Americans travel to Florida to construct a giant cannon, and then use it to launch an aluminium capsule containing a three-man crew. Around a century later, the Apollo missions lift off about a hundred miles away from the location chosen by Verne, use aluminium-based spacecraft and carry three astronauts.

– In Verne's story, the project is enormously expensive and most of the funding came from America. Once again, Verne was right on the money. The Apollo project eventually weighed in at an eye-watering $25 billion and America picked up the bill.

– Towards the end of Verne's second story, the three astronauts splash down in the Pacific Ocean and are picked up by an American Navy warship. Again, in an almost exact mirroring of Verne's story, the Apollo astronauts land in the Pacific Ocean and are recovered by a Navy vessel.

– Finally, Verne calls his cannon Columbiad, just one letter away from the Apollo 11 Command Module, Columbia.

These striking similarities were not lost on the crew of Apollo 11, the astronauts mentioning Verne's book as the crew prepared to re-enter Earth's atmosphere in July 1969.

FROM IMAGINATION TO INNOVATION

Verne's remarkable focus on the mathematics and science of space travel inspired several generations of rocket scientists.

Konstantin Tsiolkovsky is widely seen as the Russian father of rocketry. Born in 1857, Tsiolkovsky came across Verne's novels when he was sixteen years old, became hooked on space and decided to devote his life to astronautics. Throughout his career, he frequently returned to Verne for inspiration, at one point calculating that the acceleration forces generated by Verne's giant cannon would have resulted in the three humans, two dogs, six chickens and one fine cock being reduced to a single thin gel on the floor of the capsule.

The American engineer Robert Goddard also read Verne's stories as a young boy. Goddard would make notes about Verne's calculations in the margins of the novels and later credited the stories with sparking his lifelong interest in science and astronautics. Following in the footsteps of Tsiolkovsky, Goddard also became fascinated by the power needed to escape the Earth's gravitational pull and he went on to build the world's first liquid-fuelled rocket.

Similarly, Austrian physicist Hermann Oberth came across Verne's books when he was fourteen years old and also caught space fever. Like Tsiolkovsky and Goddard, Oberth became obsessed with Verne's calculations and formulae; he ended up devoting his career to studying rocketry.

Tsiolkovsky, Goddard and Oberth set the stage for arguably the most famous of Verne's protégés, Wernher von Braun.

Born in the small Prussian town of Wirsitz in 1912, von Braun came from a wealthy background.[26] Of aristocratic descent, he had ancestral links that could be traced back to France's Philip III and England's Edward III. During his childhood, von Braun came across Verne's novels and developed a tremendous passion for space exploration.

Never one to be accused of having his head in the clouds, the young von Braun adopted a hands-on approach to reaching for the stars. For instance, when he was just twelve years old, von Braun strapped several large fireworks to a small wooden cart, lit the fuse and stood back. This early attempt to reach for the Moon outperformed von Braun's wildest expectations, with the cart careering around in a blaze of fire. Unfortunately, the local police were less enthused by von Braun's valiant attempt to escape the Earth's gravitational pull and promptly took him into custody.

Von Braun eventually mastered the tricky trigonometry that underpins rocketry and, inspired by the groundbreaking work of Goddard and Oberth, set about designing and building a liquid-propelled rocket. In 1932, the Ordnance Department of the Reichswehr, the German armed forces, became interested in his research and offered to help fund his work. Von Braun accepted the offer of financial aid and, during the Second World War, helped the Nazis to develop the world's first ballistic rocket weapon, the 'V-2'. Von Braun and his team originally worked at Peenemünde in north-east Germany, on a remote Baltic Sea island, before transferring to the notorious underground Mittelwerk complex in the Harz mountains.[27] The V-2 carried a large warhead and was capable of attacking targets more than 200 miles away. It's estimated that over 8,000 people were killed in V-2 attacks, with a further 12,000 slave-labour prisoners dying in the terrible conditions at Mittelwerk.

At the end of the war, both America and the Soviet Union were keen to develop their own ballistic rockets and so recruited some of the German scientists who had played a key role in developing the V-2. Von Braun and just over a hundred of his most experienced engineers were brought to a military base in Alabama as part of a secret project that came to be known as 'Operation Paperclip' (a paperclip was placed on the file of the scientists that were chosen to enter America). He soon became one of the key players in the country's space programme.

CREATIVITY AND THE CLOSED MIND

While waiting to start work for the American government, von Braun busied himself by writing a novel about sending spacecraft to Mars. Entitled *Das Marsprojekt* (which roughly translates as 'The Mars Project'), von Braun followed in the footsteps of Verne and did his best to ensure that the technology described in the story was as scientifically accurate as possible. The book never saw the light of day, but many of von Braun's ideas formed the basis for a series of articles published in the popular magazine *Colliers* which caught the public's attention. Von Braun became a regular guest on national television and radio programmes.[28]

In the mid-1950s, von Braun teamed-up with Walt Disney and co-presented a documentary about the future of space travel. *Man in Space* was an unusual mixture of scientists talking about rocketry and Disney's light-hearted animations. For instance, to illustrate Newton's theory that for every action there is an equal and opposite reaction, a cartoon puppy sat on a slippery surface, sneezed and shot backwards. The programme attracted over 40 million viewers and led to von Braun presenting two more space-based shows for Disney.

When NASA started to study different ways of making Kennedy's ambitious Moon-based vision a reality, it was inevitable that von Braun and his rocket men were going to be key movers and shakers.[29]

In Verne's stories, the Baltimore Gun Club had tried to fire a capsule directly to the Moon and, during the Second World War, thousands of V-2 rockets had travelled straight to their targets. Similarly, one of the options initially considered by von Braun and his team involved a rocket travelling between the Earth and the Moon (known in the trade as a Direct Ascent). This scheme involved a spaceship blasting off from the Earth, zooming through space, landing on the Moon, blasting off from the lunar surface and returning home. Unfortunately, calculations quickly showed that a Direct Ascent was easier said than done, in part because it required building a gigantic spacecraft that could hold the huge amount of fuel needed to lift off from both the Earth and the Moon.

Worried that a Direct Ascent wouldn't get off the ground, von Braun and his team began to explore a modified version of the idea, known as an Earth Orbit Rendezvous. Rather than launching one gigantic rocket away from the Earth, this plan involved several smaller spacecraft, a few more modestly sized rockets, and a multi-stage plan. First, some of the rockets would be used to put the smaller spacecraft into orbit around the Earth. Second, these spacecraft would rendezvous and be bolted together to form a gigantic mother ship. Third, a final rocket would carry a vast amount of fuel up into space and transfer it to the waiting mother ship. Finally, the mother ship would journey to the Moon, land on the lunar surface, lift off and return home. An Earth Orbit Rendezvous avoided having to lift a gigantic rocket away from the Earth, but it still faced a major problem. To be successful, astronauts would have to be able to safely reverse-park their battleship-sized spacecraft on a distant world. Worse still, this landing would have to be carried out 'fins first' and with the spacecraft carrying a large amount of fuel needed to lift off from the lunar surface. Try as they might, von Braun and his fellow rocket scientists struggled to find a convincing solution to the problem.

Despite the issues involved in both a Direct Ascent and an Earth Orbit Rendezvous, Team Braun remained firm supporters of both schemes.

While von Braun was busy trying to work out how to go directly from the Earth to the Moon, a group of young engineers at NASA's Langley Research Center in Virginia were exploring a radically different approach to the problem.

'YOU MAY FEEL THAT YOU ARE DEALING WITH A CRANK'

During the 1950s, the Langley engineers had realized that they had a somewhat impoverished understanding of space exploration. Obviously, they knew how to put an aeroplane in the air and, on a good day, keep it there. However, when it came to moving a spacecraft around in the vacuum of space, none of their ideas about lift, thrust, and elevation made much sense. A young group of engineers were charged with investigating the issue and began to immerse themselves in the complexities of celestial mechanics and interplanetary travel. One of the most active members of the group was a young engineer named John Houbolt.[30]

Houbolt had grown up on a small farm in Illinois and, like von Braun, had developed an early hands-on interest in flight – when he was a young child, Houbolt opened an umbrella and hurled himself out of the hayloft on his farm).[31] Fascinated by mathematics and engineering, Houbolt went on to study technical science at college and eventually joined Langley. When Houbolt and his colleagues heard about Sputnik, they thought it likely that America would become involved in a race to the Moon and began to investigate the best way of ensuring a successful lunar landing.

Unlike von Braun, these young engineers hadn't spent the Second World War developing ballistic missiles and so didn't have a love of huge rockets travelling directly from A to B. Instead, they were more open to other approaches and, after carefully analysing a range of options, came up with a completely different plan.

Houbolt wanted to build a spacecraft that consisted of two quite different parts. The first part would house the astronauts, along with their supplies, equipment, and fuel. The second part would consist of a small Lunar Lander that was specifically designed to ferry the astronauts to, and from, the lunar surface. According to Houbolt's plan, this two-part spacecraft would blast away from the Earth, make its way through space, and orbit the Moon. The astronauts would then climb into their Lander, and head down to the lunar surface. After walking around on the Moon, the astronauts would climb back into the Lander, lift off and reconvene with their orbiting spacecraft. Finally, the astronauts would jettison the Lander and head back towards the Earth.

This plan, known as a 'Lunar Orbit Rendezvous', was clever because it avoided taking any unnecessary weight to, or from, the lunar surface. The fuel and supplies required to return to the Earth, along with the heat shield needed to ensure a safe re-entry through the Earth's atmosphere, all remained with the orbiting spacecraft. In addition, Houbolt's plan allowed engineers to create a module specifically designed for the lunar landing, and also allowed the astronauts to jettison the Lander before heading home, thus further reducing the amount of fuel needed for their trip.

Von Braun thought that getting to the Moon would probably involve landing a giant spaceship on the lunar surface. In contrast, Houbolt argued that it was much better to use a much smaller landing craft. As Houbolt liked to put it, while von Braun wanted to build a huge Cadillac, he preferred a much more modest Chevrolet.[32]

Houbolt was convinced that his scheme was far better than either a Direct Ascent or an Earth Orbit Rendezvous, and thought that convincing his fellow engineers would be easy. Unfortunately, his optimism proved seriously misplaced. In the early 1960s, Houbolt pitched to NASA administrators on several occasions – time and again the various committees rejected his scheme (remarks apparently included: 'He doesn't know what he's talking about' and 'His figures lie').[33]

Some members thought that attempting a rendezvous while orbiting the Moon was too risky. Fearing the worst (and perhaps mindful of the cliffhanger ending to Verne's first space story), they worried about the spacecraft being transformed into a high-tech coffin, endlessly circling the Moon as a very public reminder of American failure. Other committee members were simply more comfortable with the traditional solutions being proposed by von Braun and his team. Try as Houbolt might, the consensus still supported a Direct Ascent or Earth Orbit Rendezvous.

Eventually, in a last-ditch attempt to promote his scheme, Houbolt travelled to NASA's headquarters and outlined his plan to the agency's leaders. Once again, he received a less than enthusiastic response, and after a long debate, the idea was rejected.

Ever the mathematician, Houbolt decided to appeal to a higher power. He broke the chain of command, bypassed official channels and sent a letter directly to one of NASA's senior leaders. It was a brave, but potentially foolhardy move that might have cost him his job. Houbolt began by admitting that the idea of a Lunar Orbit Rendezvous was not especially popular with some of his colleagues, and that his unorthodox approach might make him appear rather eccentric:

> Somewhat as a voice in the wilderness, I would like to pass on a few thoughts that have been of deep concern to me over recent months . . . Since we have had only occasional and limited contact, and because you therefore probably do not know me very well, it is conceivable that after reading this you may feel that you are dealing with a crank. Do not be afraid of this.

Houbolt then spent the following nine pages outlining the challenges facing the Apollo programme and explaining how he believed that a Lunar Orbit Rendezvous provided an elegant, and achievable, solution. After months of additional negotiations and analysis, Houbolt's persistence paid off and the key decision makers, including von Braun, eventually saw sense and switched their allegiance to his scheme. Houbolt was on a work trip to Paris when NASA publicly announced their decision. Houbolt's boss shook his hand and congratulated him on single-handedly saving the American government billions of dollars.

Time magazine later described Houbolt as 'Apollo's unsung hero' and declared that had the decision makers stuck with the idea of a single gigantic rocket, the Apollo programme would have probably failed to achieve Kennedy's dream of reaching the Moon before the decade was out. When he was interviewed about his contribution to Apollo, Houbolt would frequently say that the proudest moment in his life came when Neil Armstrong stepped onto the surface of the Moon. By then, Houbolt had left Langley Research Center but was invited to Mission Control to witness the historic event. As Neil Armstrong set foot on the lunar surface, von Braun turned to Houbolt and said, 'John, it worked beautifully.'

*

Creating ideas that are out of this world

As the writer Zig Ziglar once memorably remarked: 'People do not wander around and then find themselves at the top of Mount Everest.'

To be successful, you need a plan. Sometimes the route ahead will be obvious and straightforward, and will therefore involve taking a well-trodden path. However, when it comes to achieving the seemingly impossible, it's often important to do what John

Houbolt did – jettison tradition and create a more innovative way of getting where you want to go.

Let's try an exercise. Take a quick look at these six street plans, and decide the shortest route from A to B on each map.

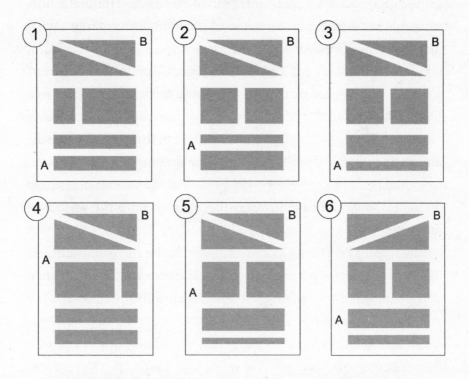

This exercise is based on research carried out by psychologists just after the Second World War.[34] As you might have noticed, there's a diagonal street on each of the plans. On the first five plans, the direction of the diagonal doesn't help you get to the end point quickly. However, on the sixth map, the direction of the diagonal has been changed, and so now it can be used as a short cut. Unfortunately, people often fail to spot this shorter route because the first five maps prime their minds to ignore diagonal roads. Researchers have labelled this curious phenomenon the Einstellung (German for 'attitude') effect.

This almost prevented the entire Apollo programme from getting off the ground. Von Braun's wartime work had involved launching missiles that were designed to travel straight to their target. In the same way that many people fail to spot the diagonal road short cut, so it's possible that von Braun's team didn't realize that their wartime experiences were partly responsible for them overlooking other pathways to the Moon.

Unfortunately, we're all prone to the Einstellung effect. From scientists to students, designers to software developers, entrepreneurs to engineers, once we have found one way of solving a problem, we often become blind to even the most obvious of alternative answers.

This type of thinking is not the only barrier to innovation. In 2016, psychologist Aiden Gregg conducted a simple but fascinating experiment involving an imaginary planet.[35] Hundreds of volunteers were asked to imagine that they were living on a distant world inhabited by creatures called Niffites and Luppites. Half of the volunteers were asked to imagine that they had come up with the theory that the Niffites were predators and the Luppites were prey. In contrast, the other volunteers were told that someone called Alex had come up with that theory. Everyone was then asked to assess evidence that was either pro or anti the theory. When the volunteers believed that they had created the theory, they stubbornly resisted any evidence suggesting that it wasn't true.

Von Braun's team liked the idea of sending a gigantic rocket to the surface of the Moon, in part, because it was *their* idea. And, just like the people imagining the Niffites and Luppites, they were therefore overly reluctant to let go of their beloved scheme. John Houbolt saved the day by overcoming these obstacles. Refusing to become trapped by received wisdom, or becoming attached to an idea simply because it was of his making, Houbolt questioned conventional thinking and began to objectively explore other solutions.

The same type of innovative thinking has changed the world. People used to communicate over long distances using telegraphy until Alexander Graham Bell invented the telephone. They took photographs using 35mm film, until an engineer at Eastman Kodak built the world's first digital camera. Almost everyone used to turn to textbooks for facts and figures until Tim Berners Lee created the World Wide Web. And families used to huddle around their television sets until three young entrepreneurs invented YouTube.

It's easy to believe that these creative movers and shakers are a world apart from the rest of us, and that they have an innate ability to develop radically new ideas and plans. But is that really the case?

Let's try a second exercise. Imagine being given a handful of paperclips and being asked to think of as many uses for them as possible. Give yourself 59 seconds and see how many ideas you can come up with.

How did you do? Maybe you thought of bending them into the words 'Happy Birthday' and using them to decorate a cake. Or perhaps you came up with the idea of linking several of the clips together to create a lovely office-themed bracelet. Or using one of the clips to hold your shirt together if you are unfortunate enough to lose a button. Or using one of the clips as a bookmark.

Or having several of them hold up the hem on a dress. Or even using the paperclips to hold pieces of paper together!

Psychologists have discovered that this seemingly simple test is a surprisingly accurate way of measuring an important form of creativity known as 'divergent thinking'. Divergent thinking involves coming up with lots of possible solutions to a problem and, as you might expect, plays a central role in innovation and invention. Over the years, researchers have developed several tests of divergent thinking and used them in a large number of experiments. Some of this work has examined the scores from thousands of children and adults to discover how creativity changes over their lifespans.[36] The results show that very young children tend to be surprisingly creative, but that this high level of performance suddenly takes a dive when they are around 9 to 10 years old (a phenomenon that American researchers have referred to as the 'fourth grade slump'). During adulthood, people tend to regain some of this lost ground, but often struggle to reach the dizzy heights exhibited during their early years. Some researchers blame this mysterious slump in performance on modern-day educational systems. According to this hypothesis, when children first go to school, they are encouraged to have fun, play and create. However, over time, children are increasingly encouraged to be critical, rather than creative thinkers. They are presented with problems that have a single solution and encouraged to find that one correct answer. In short, they are encouraged to grow up, stop playing and start conforming.

Whatever the explanation, the good news is that it is relatively easy to get in touch with your inner child and get your creative juices flowing again. In fact, you might already be moving in the right direction. As we found out in the previous chapter, the mission controllers weren't motivated by money but were instead driven by passion. This may have helped them to be creative, with research showing that paying people for ideas reduces creativity, while having a passion for what they are doing makes them more innovative.[37] If you have a genuine love for getting to your Moon then you will be more likely to think of innovative ideas and plans. And the really good news is that other research has resulted in several simple techniques to further enhance your creativity. You might think that these techniques will involve sticking an inspirational poster on your wall, donning a kaftan and trying to contact the artist within. In fact, none of those sorts of approaches are effective. Instead, it's a case of resisting the temptation to fall in love, embracing vice versa thinking, realizing that less is more, and taking it easy.

Or you could use them to make some small hangers for your children's doll clothes.

RESISTING TEMPTATION

Take a look at this list of six animals.

Dog	Snake
Dolphin	Bear
Goldfish	Tiger

Now think of the most interesting way of splitting them into two groups of three animals and jot down your solution in the box below.

My first group would consist of these three animals.	My second group would consist of these three animals.

Why do you think this the most interesting way of grouping the animals? Is it especially useful, scientifically accurate, funny, clever, creative, or innovative? Briefly jot down your justification here:

_____

_____

Right now, you are beginning to see the world in a certain way. In your mind, the way in which you grouped the animals is slowly starting to take hold and preventing you from seeing other ways of grouping them. Not only that, but you are starting to fall in love with your answer because it came from somewhere very close to your heart, namely, your head. Unfortunately, this is the kind of thinking that resulted in von Braun's team sticking with their plans about how best to get to the Moon, instead of shifting to the much better idea proposed by Houbolt.

Preventing this kind of thinking is surprisingly easy. In fact, we can do it right now. Imagine being told that your original grouping was OK, but that there's a more interesting solution out there. Please come up with another way of grouping the animals:

My first group would consist of these three animals.	My second group would consist of these three animals.

And another:

My first group would consist of these three animals.	My second group would consist of these three animals.

And another:

My first group would consist of these three animals.	My second group would consist of these three animals.

When people complete this exercise, they come up with lots of ways of grouping the animals. They might, for instance, split them into animals that could be pets and animals that are more on the wild side. Or animals that have fur and animals that don't. Or animals that are in the left hand column and animals that are in the right hand column. Or animals that have four legs, and animals that don't have four legs. Or animals that you would enjoy spending some time with and animals that you would go out of your way to avoid. Or animals that are relatively large and animals that are relatively small. Or animals that can survive under water for thirty minutes and animals that can't. Or animals that are in Rudyard Kipling's *The Jungle Book*, and animals that aren't. And so the list goes on.

When you are trying to come up with an innovative plan to achieve your aims and ambitions, don't stick with the first idea that pops into your head. Even if that idea appears to be the best thing since sliced bread, force yourself to produce several alternatives. You might have been right first time round, but before you go with your initial thought, be open to alternatives and other possibilities. As the French philosopher Émile Chartier once remarked: 'Nothing is more dangerous than an idea when it is the only one you have.'

LESS IS MORE

In 1954, *Life* magazine criticized the books that schools were using to teach children to read (*Fun with Dick and Jane*), arguing that they were both dull and encouraged double entendres. In response, a publisher challenged author Theodore Geisel (a.k.a. Dr Seuss) to select 250 words from a first grader's vocabulary list, and to use the words to create an entertaining children's book. Published in 1957, Geisel's book, *The Cat In The Hat* (which was originally supposed to be about a Queen cat, but 'Queen' wasn't in the list of words) went on to sell around a million copies.

A few years later, another publisher bet Geisel $50 that he couldn't produce a second bestseller using just 50 words. Once again, Geisel came up trumps, with *Green Eggs and Ham*, selling in excess of 8 million copies. Each time, Geisel was using a highly effective creativity tool.

People often think that the more you have, the more innovative you will be. In fact, research suggests that the opposite is true. A few years ago, Irene Scopelliti, from City, University of London, studied the effects of budget restrictions on innovation.[38] Volunteers were given a list of twenty objects and told the price of each item. Some of the volunteers were then asked to choose as many of the objects as they liked, and to use them to create a new children's toy. In contrast, the other volunteers were also allowed to choose any of the objects, but were required to stay inside a fixed budget. The first group of volunteers chose more items, but the second group produced the more innovative and interesting toys.

The 'less is more' technique is as simple as it is powerful. Let's try it right now.

First, imagine that your dream is to run an Italian restaurant. You want to make yourself stand out from the crowd by creating a new type of pizza. What would you come up with? Imagine that you only had half the amount of dough or topping usually used to make a pizza. What new kind of pizza would you create?

_____

_____

Second, imagine that you have always wanted to own an art gallery. You are eager to stage innovative and groundbreaking exhibitions but only have a tiny budget for your art exhibition. What would you do?

_____

_____

When people complete this exercise they often end up being especially creative. They might, for instance, create a vertical, cone-shaped pizza, a pizza that consists of just a circular crust, containing the topping in the centre, several tiny pizzas or one that has a ball of dough at the centre, surrounded by lots of lovely topping. And when it comes to the exhibition, they consider saving on the costs of canvases and frames by having the artists come and draw directly on the walls, think about attracting commercial sponsorship by having all the pictures feature product placement or invite the public to bring whatever art materials they want and create the exhibition themselves.

When it comes to creativity, remember that less is more. Imagine that you have half of your budget or resources or that your deadline was suddenly halved. What innovative plan would you come up with?

Moonshot Memo
Subject: Vice Versa Thinking

Houbolt's Lunar Orbit Rendezvous was almost an exact opposite of von Braun's Direct Ascent and Earth Orbit Rendezvous. Von Braun favoured using a gigantic spaceship, while Houbolt opted for a small Lunar Lander. Von Braun was fixated on using a single rocket, while Houbolt argued for a multi-part craft. And von Braun was keen on taking a straight pathway to the Moon, while Houbolt focused on rendezvous and orbits.

Houbolt's clever solution is a perfect example of a creativity technique known as 'Vice Versa Thinking'. In a nutshell, to be more creative, think about what everyone else is doing and do the opposite. If everyone is going big, go small. If everyone is going slow, go fast. And if everyone is going up, go down.

When it comes to producing an innovative plan, use Vice Versa Thinking to help you part with tradition and break new ground.

Taking It Easy

Take a look at these three puzzles.

Puzzle 1: A new type of plant has been discovered on a distant planet. It doubles in area every 24 hours. One day, astronomers see that a crater has one of these plants in the middle of it. It takes 60 days for the plant to completely cover the crater. On which day was the crater half covered?

Puzzle 2: Neil and Buzz are two friends. They get on well, but have one strange quirk. When they are in the same room, Neil insists on sitting behind Buzz, but Buzz insists on sitting behind Neil. How can this seemingly impossible situation be resolved such that both of them are happy?

Puzzle 3: Imagine that you are a medieval knight. You arrive at a castle armed with a lance that is five feet long. The guard informs you that no one is allowed in the castle with any object that is over four feet long. You are understandably upset, but then have an idea. You head into town, find the local carpenter and ask him to make something. You then return to the castle and the guard lets you in. Better still, your lance is fully func-tioning, and has not been cut in any way. How did you manage to do it?

Puzzle 4: A magician says that he can move a ping-pong ball so that it goes a short distance, comes to a complete stop and then reverses itself. The magician says that he won't bounce the ball against any object or attach anything to it. How can he perform such an amazing feat?

Now give yourself **three minutes** to try to find a solution to all four puzzles. Don't worry if you can't solve some or, indeed, any of the puzzles. OK, go!

Puzzle 1:_____
Puzzle 2:_____
Puzzle 3:_____
Puzzle 4:_____

Now it's time to take a break and find out more about these exercises. Creativity researchers have presented volunteers with these types of puzzles in hundreds of studies.[39] Some of the volunteers have been given **six minutes** to try to come up with their solutions. In contrast, others have been asked to work away for **three minutes**, then given a short break and asked to work away for another **three minutes**. As you might have figured out by now, you were in the second group of volunteers! In fact, it's now time for your second attempt to solve the puzzles.

Give yourself another **three minutes** and see how you get on.

Puzzle 1:_____
Puzzle 2:_____
Puzzle 3:_____
Puzzle 4:_____

How did you do? If you are still struggling with one or more of the puzzles, here are the solutions.

Puzzle 1: Day 59, because then the plant doubles on day 60.

Puzzle 2: Have Neil and Buzz sit back to back!

Puzzle 3: You have the carpenter make a case that measures three feet by four feet and then place your lance diagonally across the box.

Puzzle 4: The magician throws the ping-pong ball up in the air.

Even though volunteers in both groups spend the same amount of time on the puzzles, those taking the brief break outperform the others. This is known as the 'incubation' effect, and seems to be due to your brain unconsciously working away on the puzzles during the downtime. The studies have also revealed when the incubation effect is especially effective.[40]

First, you have to work on the problem before taking the break; the more you slave away, the greater the impact of the down time. Before coming up with the idea of a Lunar Orbit Rendezvous, John Houbolt and his colleagues spent months carefully considering how best to land a man on the Moon. And this was just the tip of the iceberg. During their childhoods, many of the scientists and engineers involved in Apollo were busy playing with construction toys, building model aeroplanes, creating makeshift rockets and throwing themselves out of haylofts holding umbrellas. On the face of it, such shenanigans might seem like child's play. In reality, these experiences are all part of the learning process and help provide food for unconscious thought.

Second, when it comes to taking a break, it's especially effective to spend time in a relaxing and undemanding activity. You might, for instance, follow in the footsteps of Steve Jobs, Mark Zuckerberg and Jack Dorsey and go for a walk. Psychologist Marily Oppezzo from Stanford University tested people's creativity while they were either sitting at a desk or walking on a treadmill. Participants did better while walking than sitting, increasing their creative output by an average of 60%. Not only that, but the effects had a surprisingly long lifespan, with the walkers continuing to be creative even when they had sat back down.[41]

And if strolling isn't for you, consider taking a quick nap, having a bath, engaging in a spot of meditation, sleeping on it, heading off to a flotation tank, having fun with a colouring book, zoning out or daydreaming. For Apollo's John Houbolt, inspiration could strike at any moment and so he was often found scribbling down his thoughts on any nearby surface, including shopping bags, envelopes and, on at least one occasion, the side of his bathtub.

Research supports the power of taking it easy. For instance, in one study conducted by Professor Ullrich Wagner from the University of Lübeck, volunteers were given lists of numbers and asked to replace certain digits with other digits.[42] Unbeknown to them, there was an innovative way of carrying out the tedious task, such that it became quick and easy. Some of the volunteers were asked to start working on the task in the evening, go to sleep and then resume work the next morning. In contrast, other volunteers were asked to start the task early in the morning, take a break from it during the day and then resume work in the evening. Remarkably, 60% of the volunteers who headed to bed, ended up spotting the hidden and creative way of carrying out the task, compared to only 23% of those who had remained awake.

And don't worry if you don't have time for a solid eight hours of sleep. Other research shows that people become more creative when they take a short midday nap.[43] No wonder then that several of the world's most successful organizations, such as Google, Nike and Ben and Jerry's, are now encouraging their employees to take forty winks.

When it comes to innovation, remember the incubation effect. Spend some time thinking about how to solve your problem. Try to find inspiration by talking to your friends and colleagues, searching on the Internet or reading around the topic. When you start to feel as if you have gone as far as you can go, down tools, walk away, hit the hay, take a nap and let your unconscious brain take the strain.

SUMMARY

It's often important to be able to come up with several options and to let the best idea win. And the more innovative and original the ideas, the better. Maximize this process by the following:

– Avoiding the temptation to go for the first plan that occurs to you. Force yourself to come up with several other ideas and make sure that you don't fall in love with any of them until you are sure you have met the right one.

– Utilizing Vice Versa Thinking. Identify what everyone else is doing and consider doing the opposite.

– Remembering that less is more! Imagine that you only had half of your resources, time, energy or funds. What would you do then?

– Finally, take it easy. Work on your plans and ideas for a while; then walk away. Take a break, have a bath or go to sleep. Then return to the issues and see what pops into your head.

Moonshot Memo
Subject: A cup of creativitea

Here's another quick way of promoting innovation.

Researchers from Peking University recently assembled a group of volunteers and gave half of them a cup of tea.[44] All of the volunteers were then asked to make an attractive shape from children's building blocks and to come up with a new name for a ramen noodle shop.

Another group of people then rated how creative they found each of the shapes and shop names. Remarkably, the volunteers that had drunk the cup of tea were far more innovative than their non-tea drinking colleagues.

Many researchers think that this effect might be due to the tea making the volunteers feel more relaxed and this, in turn, then opening their minds to more creative thoughts.

Either way, when you want to come up with an innovative plan, try putting the kettle on.

3

'WE DIDN'T KNOW THAT IT COULDN'T BE DONE'

Where we encounter the brilliant leader that put his
faith in a group of young people, and discover how you
can reap the rewards of self-belief.

Kennedy had assured the world that America would go to the
Moon before the decade was out. Rocket scientists had debated the
ups and downs of several mission plans and the innovative John
Houbolt had thought up a scheme that was creative and smart.
Progress was being made, but one key issued remained – who was
going to make Kennedy's vision a reality?

Spooked by the sudden appearance of Sputnik, the American
government worried that the Soviets might be close to launching
a human into space and hurriedly set up their own manned space-
flight programme. Their initial and somewhat small-scale affair
involved a handful of scientists coming together under the aus-
pices of a secret project called Man In Space Soonest (or MISS
for short).[45] In 1958, the government decided to invest far greater

resources in space exploration and formed the National Aeronautics and Space Administration (NASA).

Within months this new agency launched an ambitious programme that aimed to restore America's failing reputation by putting a human into orbit around the Earth. Entitled Project Mercury, it was clear that success would depend on finding astronauts willing to climb into a fuel-filled rocket, blast into space, withstand weightlessness and face the terrifying amount of heat generated by re-entering the Earth's atmosphere at thousands of miles per hour. After briefly considering recruiting circus acrobats, Project Mercury bosses turned their attention to test pilots.

THE RIGHT STUFF

Potential candidates had to have at least 1,500 hours of flying experience, be under 40 years old, hold a degree in engineering or an equivalent discipline and stand no taller than 5ft 11 inches (in order to fit into the small space capsule). After carefully examining the records of over 500 military personnel, a handful of hopefuls received a top secret message asking them to report to a mysterious address in Washington, D.C. Around thirty candidates made it through an initial screening interview and were invited to undergo a series of extensive physical and psychological examinations.

At the time, some scientists were concerned about the terrible effects that space exploration might have on the human body, with some predicting that the weightlessness alone could make the astronauts' eyeballs distort, prevent swallowing and cause prolonged periods of vomiting. As a result, the Project Mercury managers were keen to recruit those in tip-top physical condition and so set about exploring every inch of the candidates' bodies. As part of the process, the candidates had their eyeballs gyrated, spent hours in pitch-black acoustic isolation chambers and were spun around at tremendous speed. These highly demanding tests were interspersed with less challenging tasks, including seeing how long the candidates could keep their feet in a bowl of iced water and how many balloons they could blow up before becoming exhausted.

A second set of examinations was designed to discover whether the candidates were made of the right mental stuff.[46] These examinations attempted to assess candidates' ability to cope with stress, overcome fear and perform under pressure. Some of the tests were rather questionable, including attempts to discover whether candidates might have an unconscious death wish by asking them to describe the images that came to mind when they looked at random inkblots. Some of the candidates struggled to take the tests seriously and refused to play by the rules. In one especially memorable inkblot examination, candidate Pete Conrad was asked to look at a blank card and describe the images in his mind. After a few moments of silence, Conrad calmly informed the psychologists that the blank card was being held upside down.

Just seven candidates made the grade, including the man destined to become the first American in space, Alan Shepard. Known as the Mercury Seven, all of the astronauts were in great physical shape and deemed to be in possession of the type of highly stable minds that would enable them to keep calm in the face of impending disaster. This latter quality was frequently called into play when the astronauts encountered the latest technology to emerge from Project Mercury. In 1960, for instance, the seven astronauts attended an unmanned launch of the type of rocket that would be used to take them into space. The event was a spectacular failure, with the rocket exploding just seconds after lift-off. After watching the colossal explosion, Shepard calmly turned to his colleagues and remarked: 'Well, I'm glad they got that out of the way.'[47]

On 12 April 1961, the American space programme was dealt yet another blow when Russian cosmonaut Yuri Gagarin became the first man in space. Russia is several hours ahead of America, and at 4 a.m. a Project Mercury public affairs officer was woken up by a journalist and asked for a comment. The blurry-eyed official said that he knew nothing of Gagarin's historic flight and declared, 'We're all asleep down here.' The comment was reported by newspapers around the world and was yet another source of national embarrassment.

Less than one month later it fell to Alan Shepard and his spacecraft, Freedom 7, to restore the nation's pride. On 5 May 1961, Shepard donned his tight-fitting silver space suit, climbed into Freedom 7 and readied himself to blast away from the surface of the Earth. When reporters later asked Shepard what went through his mind as he lay on top of the giant rocket, the astronaut quipped: 'That every part of this ship was built by the lowest bidder.'[48]

After several long delays, Shepard blasted away from the Earth and was subjected to acceleration forces in excess of 6 G (causing his body to feel as if it weighed six times more than normal). Once in space, Shepard was only able to survey his surroundings via a small periscope. During the launch delays, he had placed a grey filter over the periscope's lens to prevent himself being dazzled by sunlight. Unfortunately, he had forgotten to remove the filter and so when America's first astronaut peered through the periscope, he could see very little.[49] Eager not to disappoint the millions of people following his flight live on television, Shepard looked at the strange grey blob in front of him and remarked: 'What a beautiful view.'

The flight lasted just fifteen minutes and was a simple 'up and down' suborbital flight. After a bumpy re-entry through the atmosphere, Freedom 7 splashed safely down and newspapers across the world carried photographs of the blue-eyed astronaut across their front pages – *Newsday*: 'Our Shep does it', *Daily News*: 'What a ride!', *Chicago Daily News*: 'Space Hop A Success.'

CREATING THE 'LEADERSHIP LABORATORY'

When it comes to space exploration, brave astronauts like Shepard rightly take centre stage. However, dig a little deeper into the mission archives and a second group of people quickly emerges. These people avoided all of the glitz and glamour and were instead content to stand quietly in the shadows. They are not widely associated with putting a person on the Moon and yet were central to the success of the entire enterprise. And they exist because of the clever thinking and hard work of the remarkable Chris Kraft.

Born in Phoebus, Virginia, in 1924, Chris Kraft came from a modest background.[50] At the time, Phoebus was a tough railroad town. The local dump bordered directly onto his backyard and some of Kraft's earliest memories involved watching workers incinerating huge piles of trash and seeing the large plumes of smoke climb high into the sky. Hard work was a way of life in Phoebus and when he wasn't in school, Kraft was busy unloading freight trains and serving in local stores.

In his wonderful autobiography, *Flight*, Kraft describes being inspired by his high-school mathematics teacher ('Just think about what you want in your life and you can do it'), and coming to realize that there was more to life. Kraft went on to obtain a degree in aeronautical engineering from Virginia Tech; he then accepted a position at Langley Field, researching aviation technology. In the late 1950s, the American government wanted to encourage some of the nation's aviation experts to turn their attention to space technology. Kraft stepped up to the mark and devoted himself to restoring the nation's failing reputation in space. Kraft would later joke that his full name – Christopher Columbus Kraft Junior – perhaps meant that he was destined for a lifetime of exploration.

Kraft quickly worked his way up as the American space pro-gramme started to get off the ground. In January 1961, Kraft sent a rocket containing a chimpanzee called Ham (named after the laboratory that had looked after him – the Holloman Aero-space Medical Center) on a sixteen-minute suborbital flight. Ham's 'chimponaut' mission proved successful and gave the green light to sending Shepard into space. When Shepard heard the news of his forthcoming flight, he allegedly quipped: 'I guess they ran out of monkeys.'[51] Almost right from the start of the entire enterprise, Kraft realized that the success of the missions depended on having ground crew who could quickly provide astronauts with real-time monitoring and support. Kraft also realized that all of these people needed to be housed in the same room, alongside the monitors, consoles and screens that were vital to their work. For the early Project Mercury flights this room was a relatively small-scale affair and based at Florida's Cape Canaveral. In 1965, the room was expanded and came to be known as Mission Control; it was then moved to the Manned Spacecraft Center in Houston, Texas. Over the years this now iconic space has had many nicknames, including 'The Cathedral', 'The Palace' and 'The Leadership Laboratory'.

Kraft needed to populate Mission Control with people who were able to make the seemingly impossible possible. The Mercury astronauts had endured an extremely challenging, tough and long-winded selection procedure. They had been subjected to incredible forces, sat in bowls of freezing water and been asked to interpret endless inkblots. The handful of astronauts that made the grade were all drawn from the military, were physically fit, able to carry out their duties under extreme pressure, in their late thirties and married with children. In general, they came from comfortable, middle-class backgrounds. When it came to hiring mission con-trollers, Kraft had a very different type of candidate in mind.

THE OTHER RIGHT STUFF

Kraft knew that the road ahead was going to be tough, but he had faced adversity in the past and had managed to find a way through. Perhaps eager to create a team in his own image, Kraft was drawn to people who came from modest backgrounds and had succeeded in making something of their lives. People who had worked hard and made their own luck. People who could imagine a brighter future and had what it took to make change happen. Many of the people chosen by Kraft came from rural and farming backgrounds and were frequently the first in their families to go to college. They had a high work ethic, a passion for space exploration and an appetite for taking on tough challenges.

Perhaps most important of all, they were an amazingly young bunch who were willing to develop, grow and learn. Indeed, when Neil Armstrong walked on the Moon, the average age of the people in Mission Control was an astonishing twenty-six years old.

Jerry Bostick was typical of the people selected by Kraft. Bostick grew up on a small family farm in rural Mississippi and as a young child often worked from sunrise to sunset, helping to grow cotton and corn.[52] He soon adopted his family's strong work ethic: 'No matter your job, make sure that you do it better than anyone else'; 'You can accomplish anything, but you have to work hard.' In his teens, he added to his daily farming chores by delivering newspapers, working as a petrol pump attendant and selling popcorn in his local cinema at night.

Bostick served as a Page and Doorman in the US House of Representatives, graduating Capitol Page School as Valedictorian. He attended Mississippi State University, graduating with a degree in Civil Engineering, and then started to search for a job. Bostick eventually ended up working at the NASA Langley Research Center, but was disappointed to find himself assigned to projects that were more blue-sky thinking than practical problem solving.

Like many engineers at the time, Bostick was energized by Kennedy's vision of putting a man on the Moon, and so applied to work on more space-related projects. Unfortunately, at his interview he was told that the focus was on recruiting aeronautical experts, not civil engineers. However, on his way out of the interview room, Bostick happened to bump into Kraft and the two of them started to chat. After a few moments, Kraft made a snap decision and told the recruiting officer to offer Bostick a position – 'Hell, hire him. We may need somebody to survey the Moon.' Bostick slowly worked his way through the ranks and was eventually put in charge of the group that helped ensure that the Apollo spacecraft headed in the right direction, known as Mission Control's Flight Dynamics Branch. He was just 29 years old when he helped make Kennedy's dream a reality. Looking back, he realized that the young mission controllers were an optimistic group:

They decided to go with a bunch of young guys fresh out of college because we didn't know that it couldn't be done! When we were told that we needed to find a way of getting to the Moon, we just got on and did it. I can remember thinking 'Man alive, that's a stretch', but the President had set out a really clear goal in a single sentence, and it was up to us to make it happen.[53]

Kraft's recruitment drive paid off and Project Mercury went from strength to strength. In July 1961, astronaut Gus Grissom followed in the footsteps of Alan Shepard, piloting a second suborbital flight into space. In February 1962, John Glenn became the first American to enter orbit, spending nearly five hours zooming around the Earth. Over the next few months, the astronauts spent even more time in space, with Gordon 'Gordo' Cooper eventually setting a new American record by orbiting the Earth for around a day and a half. Although the project was highly successful, many of the missions were far from plain sailing. During Cooper's nineteenth orbit, for instance, his urine collection system leaked, resulting in the interior of the capsule being covered in a fine spray of unwanted moisture. This caused several systems to short-circuit and shut down, which resulted in a dangerous rise in both air temperature and carbon dioxide levels. Cooper was forced to attempt a manual re-entry, relying on his wristwatch and knowledge of star formations to guide his descent. Remarkably, he landed less than four miles away from the Navy vessel that was due to pick him up, setting a new record for splashdown accuracy.

Project Mercury ended in 1963 and attention shifted to a second programme that aimed to send two-astronaut crews into space. Entitled Project Gemini (as in the astrological twins), ten crews flew in orbit around the Earth between 1964 and 1966, with mission highlights including several space walks, rendezvous and dockings.

Within a few short years, the American space programme had made its first faltering steps towards the Moon. Much of this success can be attributed to the bravery of the astronauts and the skill of the thousands of engineers and scientists working on the project. However, right at the centre of it all sat Chris Kraft, both instrumental in creating the concept of Mission Control and populating it with a remarkable team of people. People who came from modest backgrounds, and who were used to working hard and overcoming adversity. People who were passionate about putting a man on the Moon and were so young they didn't know that the task was almost impossible. And people who were about to face the hardest challenge of their lives.

Moonshot Memo
Subject: Tragic news

On 21 November 1963, President Kennedy visited Houston's Manned Spacecraft Centre and spoke about the day when America would launch the world's largest rocket into space. In a slight slip, Kennedy announced that the rocket would be 'lifting the heaviest payroll . . . that is payload'. The President paused, and then added, 'it will be the heaviest payroll, too!'

The following day the President was assassinated while riding in a motorcade through Dealey Plaza in Dallas, Texas. Kennedy's death sent a shock wave through the nation. However, his vision provided a beacon that continued to inspire Mission Control. As Jerry Bostick commented:

'After Kennedy was assassinated, we doubled our efforts, and became even more determined to meet his goal. We didn't sit around and talk about it, but it was very important to us. We all knew what we needed to do. We knew that we needed to land on the moon.'[54]

Had the Apollo project relied solely on Kennedy's leadership it might have been in trouble. Instead, it grew from his remarkable vision, and so marched on.

The episode illustrates another upside of a project being driven by passion – this enthusiasm continues to motivate people, even when leaders and visionaries move on, retire, or pass away.

*

Reaping the Rewards of Self-Belief

'Whether you think that you can or you can't,
you're usually right.'

– Henry Ford

Let's start with a quick thought experiment involving a geometrical puzzle. Take a look at this plan of a launchpad.

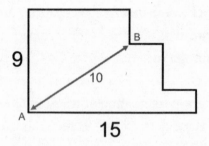

As you can see, one side of the launchpad is 9 metres long and another side is 15 metres long, and the distance between point A and point B is 10 metres. You have to buy some fencing to place all round the perimeter of the launchpad. Do you think that you have what it takes to figure out exactly how much fencing you will require? Before you make your decision, you should know that you aren't allowed to use a ruler, consult a book or ask a friend. Oh, and you will only have 2 minutes to complete the task. Do you think that you will be able to come up with the correct answer?

We will return to the thought experiment in a moment. First, let's meet the legendary psychologist Albert Bandura.

Born in 1925, Albert Bandura grew up in a tiny town in northern Canada. Coming from a large family of modest means, he quickly realized that success was only going to come through hard work, self-reliance and perseverance. When Bandura finished high school, his parents encouraged him to hit the road and see the world. Taking their advice literally, Bandura managed to get a summer job filling holes on the Alaska Highway. While digging away, Bandura noticed his fellow labourers' hardcore drinking and gambling habits, and started to develop a deep interest in matters of the mind. Over the next few years, Bandura obtained various qualifications in psychology and was eventually offered a position at Stanford University. Once there, he spent over twenty years exploring the science of success, and our thought experiment is based on his groundbreaking work.

A few moments ago you were asked whether you thought that you would be able to come up with a solution to a geometrical puzzle.

Some people are pessimistic about their chances of success. If this is the case, they often don't even try to solve the puzzle and, even if they do make a start, they give up the moment the going gets tough. As a result, their initial pessimism is transformed into a self-fulfilling prophecy, and they end up failing. As Kennedy memorably said when he addressed Congress about going to the Moon: 'While we cannot guarantee that we shall one day be first, we can guarantee that any failure to make this effort will make us last.'

The opposite is true of those that are more optimistic of success. Fuelled by their self-belief, they are willing to make a start, and are far more likely to persevere. In doing so, they increase the likelihood of discovering that challenges are not as tricky as they first seem or uncover innovative ways of achieving success. For instance, when trying to solve the geometrical challenge, the optimists might have worked away and quickly noticed that there's a remarkably easy solution. You can ignore the distance between point A and point B – that was just a distraction. Then, all you have to do is imagine moving these two horizontal lines up like this:

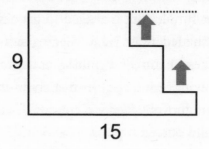

And these two vertical lines to the right like this:

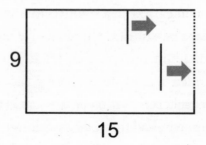

Then you are left with a perfect rectangle like this:

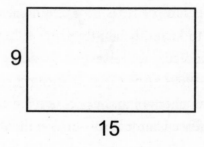

And now it's obvious that the launchpad's perimeter is 48 metres (2 sides that are 9 metres long plus 2 sides that are 15 metres long). You don't need to know any geometry at all – almost everyone solves the puzzle as long as they make a start and spend some time thinking about it.

This was the simple, but powerful, idea that fascinated Bandura.[55] He speculated that when people believe that they don't have what it takes to complete a task successfully (a state that he referred to as having 'low self-efficacy'), they would see little point in making an initial effort. Moreover, even if they did manage to get going, they would tend to quit when they encountered obstacles. Exactly the opposite is true of those that think they can. When people think that they are made of the right stuff and so expect to do well (referred to as 'high self-efficacy'), they are far more likely to make a start, persevere when they encounter bumps in the road and discover new ways of navigating the path ahead.

As a result, Bandura predicted that people's beliefs about their future performance would become self-fulfilling prophecies in pretty much all aspects of everyday life. The findings from thousands of studies have shown that Bandura was correct.[56] Whether it's patients trying to recover from illness, athletes trying to outperform their competitors, students trying to obtain top marks, activists trying to change the world, managers trying to increase the bottom line, entrepreneurs trying to create new businesses or smokers trying to kick the habit, pessimists tend to fail and optimists tend to succeed.

In the early 1960s, America's space programme consisted of several failed launches, a handful of monkey-based missions and Alan Shepard's fifteen-minute sub-orbital flight. Yet Kennedy had proudly promised the world that America would put a person on the Moon before the decade was out. Perhaps not surprisingly, many people believed that the goal was impossible to achieve. In doing so, they created the biggest obstacle of all. Chris Kraft recruited people who had a track record of overcoming adversity and believed in themselves. People that were optimistic of success and willing to give it their best shot. People who were so young that, in the words of Jerry Bostick, they didn't know that they couldn't do it.

The good news is that enhancing your self-belief is surprisingly easy, and just involves creating small wins, talking to yourself, looking in the rose-tinted rear view mirror, and indulging in a spot of hero-worship.

CREATING SMALL WINS

Professor Teresa Amabile is Director of Research at Harvard Business School, where she studies the secret lives of organizations.[57] A few years ago she asked almost 300 people from 7 companies to send her a daily email describing the day's events along with their mood, motivation and productivity. Four months, and almost 12,000 emails, later, Amabile was able to identify the factors associated with success and achievement. Some of the companies invented household gadgets, one supplied cleaning services and another ran complex computer systems in the hotel industry. Regardless, when it came to predicting success, one factor emerged time and again and can be summed up in just two words: small wins.

Amabile discovered that minor milestones often resulted in a surprisingly positive effect. People are often overwhelmed when they are faced with a seemingly impossible goal. However, when they break the goal into smaller steps, it suddenly seems more achievable and their confidence soars. In addition, each time one of these stages is accomplished, their confidence and optimism grows and this then acts as a catalyst for future success. Psychologists now refer to this as the 'progress principle', and hundreds of studies involving thousands of people have revealed that breaking down an ambitious vision into a series of small steps leads to higher performance *90% of the time*. Big achievements matter, but small steps matter more.

The progress principle helped put a person on the Moon. When Kennedy first announced his ambitious vision, everyone was astonished at the idea of sending astronauts 380,000 kilometres into space, having them walk on a distant land and then bringing them back safely to the Earth. However, when the rocket scientists and engineers identified each of the stages involved in the scheme, the task seemed more manageable. The team realized that they would need to build a giant rocket that could escape the Earth's gravitational pull, to construct a craft that could make its way through space and orbit the Moon, to design a landing vehicle that could ferry astronauts between the spacecraft and the lunar surface, and to figure out a way of allowing the Lander to rendezvous with the orbiting spacecraft. As both Project Mercury and Project Gemini began to achieve these sub-goals, everyone's confidence started to grow. By the time of the Moon landings, this confidence was at an all-time high, with engineer Jay Honeycutt noting:

'By then we thought that we were damned invincible. We all thought that we've got to do this and nobody is going to stop us.

There was no chance that we're going to fail. We don't fail. During much of the Apollo program it seemed like everybody had that attitude.'[58]

This, in turn, motivated them to move onwards and upwards towards their ultimate goal.

When you are faced with a seemingly overwhelming goal, break it into bite-sized chunks. In addition, to help ensure that the principle is especially effective, use SMartT thinking to make each of the mini-goals as Specific, Measurable and Time-constrained as possible. For instance, if you want to lose 20 pounds, maybe aim to lose 1 pound each week. Or if you are trying to create a new start-up with a six-figure turnover, aim to secure a new client each month.

And don't be afraid to move forward by looking back, with research showing that people who persevere when the going gets tough are especially likely to think about what they have already accomplished.[59]

Finally, make sure that you maximize motivation and confidence by celebrating when you achieve each mini-goal. Mission Control lit a cigar after a successful splashdown and threw parties when they had achieved another milestone. You don't have to be that extreme; even treating yourself to some chocolate cake will help (unless, that is, you are trying to lose weight).

HOW TO TALK TO YOURSELF

At the turn of the twentieth century, a somewhat strange children's short story started to appear in publications around the world.[60] Although the precise details of the story varied from one version to the next, the general plot was always the same. In the story, a large number of heavy freight cars needed to be pulled over a high hill. After several larger engines have refused the job, a much smaller engine agrees to give it a go. Constantly repeating the phrase: 'I think I can, I think I can', the little engine puffs away and eventually manages to pull the freight over the hill, proudly declaring, 'I thought I could.' For over a century, the story of 'The Little Engine That Could' has encouraged youngsters to take on tough challenges and believe in themselves.

Although most people aren't asked to help haul heavy freight over a steep hill, the vast majority of them do engage in self-talk. Unfortunately, these inner monologues often don't involve thinking, 'I think I can, I think I can', but rather 'I suspect I can't', 'I probably won't' and 'I never do.' However, the good news is that it's relatively simple to stop talking to yourself like 'The Little Engine That Couldn't', and instead become 'The Little Engine That Could'.

Let's try another thought experiment. Imagine that you are going to embark on a new venture, but can hear your inner overly critical self slowly chipping away at your confidence. In our thought experiment, maybe you are thinking of starting a new relationship ('I always fail'), launching your own business ('I know self-employment is really tough, I could never make it work') or changing career ('I have a safe job, putting that at risk would be a disaster').

If that does sound like you, try arguing with yourself. Imagine that your best friend is considering embarking on the same venture and expresses the same doubts. What would you say to them? Obviously, you would never encourage your friend to be reckless, but neither would you stand by while they make such sweeping and damning predictions about the future. Instead, you would probably find a way of being realistic and supportive. You might, for instance, challenge any overly negative assessments by pointing out the upside of past failures ('Your relationships have been tough, but you are good at learning from the past and moving on'), ask for evidence ('How tough is self-employment? Have you spoken to anyone about it?'), point out that the future might be brighter than they think ('Your job might be safe, but I can imagine you finding something that makes you much happier'), and generally try to boost their confidence ('You are a strong person, and can probably cope with whatever happens').

This 'best friend' dialogue tells you how you should be talking to yourself. Rather than accepting your pessimistic view of the future, it encourages you to be far more realistic, supportive and kind.

The same approach can also help insulate you from other peoples' overly critical comments. The seventeenth-century English poet John Donne famously noted: 'No man is an island.' Psychological research has shown that Donne is correct.[61] In the same way that people can pass on their runny nose and sore throat, so their emotion and attitudes have the power to infect you too. Spend a few moments in the company of an encouraging and supportive optimist, and you suddenly feel more passionate and positive. Similarly, hang out with someone who expresses a more pessimistic view of your abilities and future, and very soon the future will appear dull and dreary.

If you encounter someone telling you that you're destined to fail, adopt the best friend approach. Ask yourself whether they really have a point or are somehow making themselves feel good by putting you down. Do they really have your best interests at heart or is there another agenda at play? Are they always emphasizing your failures and dismissing your achievements? Rather than accepting and internalizing the criticisms, carefully consider them and think about what your best friend would say. And if the same person is repeatedly predicting that you will always fail, remember the old adage of you being the average of the five people that you spend most time with, and think about whether you might be better off surrounding yourself with more positive people.

Moonshot Memo
Subject: Good luck

Astronauts and rocket scientists are sometimes a surprisingly superstitious bunch. Before blasting into space, NASA astronauts traditionally have the same type of eggs-based breakfast that Alan Shepard consumed before his pioneering Freedom 7 flight. Similarly, as Russian cosmonauts head out to their rocket, they often urinate on the rear wheel of their transfer bus, allegedly because Yuri Gagarin relieved himself in the same spot before his historic mission. And scientists at NASA's Jet Propulsion Laboratory (JPL) often keep a jar of peanuts around as a good luck charm. Why? Because in the 1960s, the JPL's first six unmanned flights to the Moon ended in failure, but on the seventh flight a controller passed around a jar of peanuts and the mission was a success.

Although it is easy to dismiss such behaviour as superstitious nonsense, there may be more going on than meets the eye. A few years ago, psychologist Lysann Damisch had volunteers carry out several tasks, such as trying to putt golf balls into a hole and solve tricky anagrams.[62] Some of the volunteers were told that their golf ball was lucky or were asked to carry out a superstitious ritual such as crossing their fingers. These lucky charms and rituals had a dramatic effect, and resulted in the volunteers being more successful when it came to both putting and solving anagrams. The study also revealed that this success was due to the charms and rituals making the volunteers feel more confident and that this, in turn, caused them to set higher personal goals and to persist longer.

So, the next time that you feel like a little boost of confidence, don't be shy about reaching for your lucky socks, crossing your fingers, having eggs for breakfast or urinating on a nearby bus.

Look in the Rose-Tinted Rear-View Mirror

When people try to predict the future, they often think about the past. Let's imagine that you are going to try to bake a lovely cake on Sunday. How do you think your cake baking will go? If your recent burnt offerings were less than impressive, then you will probably anticipate another culinary disaster. However, if you have consistently managed to produce one great sponge after another, then you will be confident of doing it again. The same principle applies to almost every aspect of your life.

As we have seen, many of the mission controllers came from modest backgrounds. They had found ways of overcoming adversity, and managed to be successful even though the odds were often stacked against them. Over time, this resulted in them being optimistic about any new challenges that might come their way, even if it involved something as daring as sending astronauts to the Moon.

To develop this mindset, think about a time when the odds were against you, and yet you pulled through and were successful. Maybe it was when you did well in a difficult exam, won a competition when you were the underdog or successfully completed a chal-lenging project at work. Think about what you accomplished and, more importantly, how you accomplished it. Can you remember having doubts about doing well? Were others sceptical? What were the factors that caused you to be successful?

Now, whatever it is, replay the scenario in your mind. When an athlete does well during a big sporting event, television networks often replay the moment again and again. Try to do the same with your past. Imagine that you are showing this moment from your life on a huge screen to an entire stadium.

Finally, think about ways of regularly reminding yourself about this scenario. Could you perhaps put a photograph of it in your drawer so that you see it every time you open the drawer? Or maybe put a certificate on your wall or some kind of memento on your desk? Either way, use the event to remind yourself that you succeeded in the past and so there is every reason to be more confident going forwards.

Hero Worship

Being optimistic about the future isn't all about you. When it comes to boosting your belief that everything will come up smelling of roses, other people are equally important. When you see someone succeed, it shows that the seemingly impossible is indeed possible.

Find a hero. Maybe it will be someone like the American activist and lecturer Helen Keller. Born in 1880, Keller was a sickly child; she eventually lost both her sight and hearing. Determined not to let her disabilities limit her life, she slowly learned how to communicate, eventually wrote several books and became the first deaf-blind person to receive a Bachelor of Arts degree. Keller was also an active campaigner and toured America promoting the rights of women and workers. A striking illustration of how people can conquer adversity, Helen Keller once noted: 'Optimism is the faith that leads to achievement. Nothing can be done without hope and confidence.'[63]

Or perhaps you will go with an athlete, like British athlete Roger Bannister. In the early 1950s, scientists thought that it was impossible to run a mile in under four minutes. Bannister was convinced that he could prove the experts wrong. In May 1954, after months of training, Bannister finally achieved his goal by running a mile in 3 minutes 59.4 seconds. Inspired by Bannister's record-breaking performance, other runners began to follow in his (impressively fast) footsteps. By 1957, the time was down to 3 minutes 57 seconds and, in 1958, it was 3 minutes 54 seconds. Reflecting on his achievement, Bannister noted: 'However ordinary each of us may seem, we are all in some way special, and can do things that are extraordinary, perhaps until then . . . even thought impossible.'[64]

Or maybe you will choose a celebrity like Oprah Winfrey. Born into poverty in rural Mississippi, Winfrey had a challenging upbringing. Winfrey became pregnant when she was just 14 years old but suffered a miscarriage. She focused on her schooling and was given a scholarship to study communications at Tennessee State University. Winfrey was then hired to work as a news anchor on a prestigious Baltimore television station, but struggled and was essentially demoted and asked to present a lower-profile programme. Undeterred, Winfrey went on to host a talk show in

Chicago and started to attract an increasingly large audience. The show was eventually renamed *The Oprah Winfrey Show* and syndicated across the country, helping to make her one of the richest self-made women in American history.

Or your hero could be a writer like J. K. Rowling. In her mid-twenties, Rowling had separated from her husband, was unemployed, struggling to raise her daughter on her own and living in a tiny flat in Edinburgh. During a train journey, Rowling had the idea of writing a story about a young boy with magical powers. Undeterred by her difficult circumstances, Rowling scribbled away in local cafés, only to see the resulting manuscript rejected by publishers time and time again. Rowling persevered and the subsequent string of Harry Potter books has made her one of the wealthiest women in the world.

If you are trying to take a team to new heights, perhaps you would rather focus on organizations than individuals. If that is the case then you might, for instance, want to take a look at Southwest Airlines. In the late 1990s, the airline was in a financial tailspin and its managers were asked safely to cut the 40-minute turnaround time for short haul planes to just 10 minutes. Moreover, Southwest managers were given just 2 years to find ways of making and implementing the change. Aviation experts considered it an impossible goal. However, their managers studied the techniques used by Formula 1 pit crews to achieve super-fast turnaround times and carefully applied the lessons learned to their own planes. Southwest Airlines achieved their goal and profits soared. Their innovative approach has since been adopted by other short haul airlines around the world.

Alternatively, you might like to find inspiration in the way in which Japan transformed high-speed travel. After the Second World War, Japan was desperate to grow its economy. Every day, goods and people were crammed onto the train between Tokyo and Osaka, but the line was badly outdated and the 300-mile trip often took more than 20 hours. Japanese transport officials urged the nation's engineers to invent a faster train, and within a few months the team built a prototype capable of an impressive 65 miles per hour. Even though this was one of the fastest passenger trains in the world at the time, the officials set an audacious stretch goal, demanding that within 10 years the engineers create a train that could reach 120 miles per hour. Many engineers believed that this was impossible, but slowly they worked at redesigning the entire railway system and stock. As a result, the world's first bullet train was launched in 1964, achieving an astonishing average speed of 120 miles per hour. This remarkable feat of Japanese engineering transformed train travel around the globe, and within a few years there were similar high-speed railways in France, Germany and Australia.

Of course, you could just be inspired by those responsible for humanity's greatest achievement – putting a person on the Moon.

Wherever you find your hero, use their inspirational story to remind you that many people have achieved the seemingly impossible. Remind yourself of how they found the confidence and optimism to struggle through the difficult years, and eventually changed the world. Perhaps place a photograph of them in your wallet or purse or on your refrigerator door, desk or noticeboard. And, each time you see it, remember how they achieved the impossible and use their story to fuel your own success.

SUMMARY

Believing that you have the skills to achieve a goal helps you to get going, and keep going, and so dramatically increases your chances of success. To boost self-belief:

– Remember the magic of small wins. Break your big goal into smaller stages and celebrate after you achieve each of these important steps.

– Don't put up with negative self-talk. Use the 'best friend' technique to create a much more productive and positive internal dialogue and surround yourself with supportive people.

– Celebrate your past achievements. Make a note of them. Play them on the big screen in your mind. Remember that you did it before and that you can do it again.

– Find a hero, a person or organization achieving the seemingly impossible. Regularly remind yourself about their inspirational story and know that if they did it, you can do it too.

Moonshot Memo
Subject: The 'Do Nothing' option

During his early years, Kraft quickly learned important lessons the hard way. At one unmanned launch in 1960, for instance, Kraft and his team were initially delighted to see a test rocket start to take off amid a massive plume of smoke. However, when the smoke cleared, the team was less than enthused to see that the rocket was still sitting on the ground. Worse still, the capsule's parachutes, originally intended to allow the craft to return to the Earth safely, suddenly deployed and hung down the sides of the rocket. If the wind had caught the parachutes, they would have stood a good chance of bringing the entire rocket crashing to the ground.

Facing an unstable rocket filled with a vast amount of highly explosive fuel, an engineer panicked and suggested emptying the fuel tanks by using a rifle to shoot holes in them. Understandably, Kraft was less than impressed with the idea and eventually decided that the best way forward was to do nothing. The heat from the sun caused some of the fuel to evaporate and a drop in the wind meant that the rocket remained upright. Eventually the rocket became safe enough for a brave engineer to crawl inside and disarm it.

The successful resolution was a great illustration of one of Kraft's most important rules of flight control: 'When you don't know what to do, don't do anything.' Kraft's mantra is as simple as it is powerful. In an emergency, people often feel the need to spring into action. However, if they haven't fully thought through the situation they can easily make matters worse. When making an important decision, put the 'do nothing' option on the table. Ask yourself whether you have the information that you need to make an informed choice and what would happen if you chose inaction over action.

Research also shows that even if you chose to do something rather than nothing, merely introducing the 'do nothing' option may increase your motivation. A few years ago, Rom Schrift from the University of Pennsylvania ran an experiment in which volunteers were asked to tackle a word-search puzzle.[65] The volunteers were paid for each word that they could find and were free to quit at any time. Some of the volunteers were given the option of looking for the names of either famous actors or capital cities. In contrast, a second group of volunteers were offered an additional option, namely, to do nothing at all. All of the volunteers chose to tackle the puzzle, but the small change in options had a dramatic effect, with those that had been presented with the 'do nothing' option working on the task for around 40% longer and being more successful.

Schrift thinks that the effect is due to people thinking, 'I rejected the idea of doing nothing and so my chosen option must be better than that, and so I should keep on going.'

Whatever the explanation, this simple but powerful idea could be used to help encourage people to stick to a diet, complete a drug regimen, visit the gym and be motivated in the workplace.

4

A ROUGH ROAD LEADS TO THE STARS

*Where we learn about the tragic fire that forced everyone
to pull up and take stock, and find out how you can
learn from failure.*

By the mid-1960s, the Apollo team had accomplished many of
their key objectives, including learning how to reliably launch
rockets into space, have spacecraft rendezvous with one another
and keep astronauts alive in orbit around the Earth. To many, it
appeared that the tide had turned and that America might soon
pull ahead of the Soviets in the race for space. Unfortunately, the
road ahead was to prove far rockier than anyone had anticipated,
and the journey would centre around the remarkable Gus Grissom.

In 1958, Air Force pilot Gus Grissom received one of the Project Mercury top secret messages inviting him to a mysterious
meeting in Washington DC.[66] After passing the initial interview
with flying colours, Grissom underwent the extensive physical and
psychological testing. All went well until the doctors discovered
that he suffered from hay fever and they threatened to reject him.

Fortunately, the quick-thinking Grissom managed to allay their concerns by calmly explaining that, as far as he was aware, there wasn't any pollen in space.[67] Grissom made the grade and, in April 1959, was chosen to be one of the Mercury Seven astronauts.

One of the most ambitious and driven of the group, Grissom was eventually chosen to follow in the footsteps of Alan Shepard and become the second American in space. On 21 July 1961, Grissom donned his spacesuit and climbed into his spacecraft, Liberty Bell 7. In his pocket were a hundred dimes that he intended to give away as souvenirs after the flight. The launch proceeded like clockwork and soon Grissom felt the rockets ignite and his capsule blast away. All went well with the fifteen-minute sub-orbital flight and Liberty Bell 7 re-entered the Earth's atmosphere and deployed its parachutes exactly as planned. When Grissom's craft splashed down in the Atlantic Ocean, it seemed like the perfect end to a flawless mission. In reality, it was just the start of a potential catastrophe.

Whereas the hatch on Alan Shepard's Freedom 7 capsule was opened with a latch, Liberty Bell 7's hatch was to be blown off via a small explosion. Although this new design was intended to help Grissom exit the capsule, it would result in him almost losing his life. According to the mission plan, Grissom was to wait in Liberty Bell 7 until a helicopter arrived and the rescue team had attached a cable to his capsule. Then, the team would detonate the hatch's explosive bolts, allowing Grissom to make a safe exit and be airlifted up to the helicopter. Unfortunately, the highly honed plan soon started to fall apart.

Moments after the helicopter arrived on the scene, Liberty Bell 7's hatch suddenly blew open and all hell let loose. The unsecured capsule rocked around in the high waves and seawater started to flood in through the open hatch. Grissom was forced to make a quick escape and dive into the freezing ocean. Desperate to try to prevent the Liberty Bell 7 from sinking, the recovery helicopter flew as low as possible and attempted to connect a cable to the capsule. With the helicopter's wheels dipping dangerously in the water, the crew managed to attach the cable, only to discover that the waterlogged capsule now weighed far more than the helicopter could handle. Each time the helicopter attempted to lift up Liberty Bell 7, the ocean swell filled the capsule with seawater. The repeated attempts to lift the overly heavy load took its toll on the helicopter's engines and the rescue crew was eventually forced to release the $2 million capsule. Within moments, the Liberty Bell 7 disappeared below the water and made its way to the bottom of the Atlantic.

Meanwhile, Grissom was in real trouble. In his rush to get out of the capsule, he had left an oxygen valve on his spacesuit open, and the seawater was now seeping in and severely reducing his buoyancy. Weighed down by the waterlogged suit and the souvenir coins, Grissom struggled to stay afloat. Worse still, the helicopter attempting to lift the Liberty Bell 7 was churning up the sea. Moments away from drowning, a second helicopter spotted Grissom, realized what was happening and lifted him to safety.

Unfortunately for America's second man in space, the trouble was far from over.[68] Grissom often felt uncomfortable speaking with the press: on one occasion, he had even donned a large straw hat and sunglasses in an attempt to sneak past journalists unnoticed. Some reporters were less than delighted with his curt delivery and nicknamed him 'Gloomy Gus' and 'The Great Stone

Face'. At the press conferences following his historic flight, journalists often refused to focus on the success of Grissom's mission, instead repeatedly asking him whether he was responsible for blowing the hatch open early and thereby losing the multi-million dollar capsule. Grissom always refuted the accusations and the other Mercury astronauts supported his position. Later, a review board would conclude that Grissom wasn't responsible for the premature detonation of the hatch or the loss of the Liberty Bell 7. Nevertheless, to some, Grissom's historic mission felt tarnished.

Grissom had managed to cheat death. Unfortunately, on another occasion he would not be so fortunate and the tragic episode would cause a radical change of mind in those working on the space programme.

TRAGEDY, RISK AND RECKLESSNESS

Grissom's next flight was to prove a less dramatic, but equally contentious, affair. This time, the mission involved his teaming up with fellow astronaut John Young to pilot the first two-person Gemini capsule into space. In March 1965, the two astronauts blasted away from Cape Canaveral and were soon orbiting the Earth. Part way through the mission, Young told Grissom that he had managed to smuggle aboard some interesting food and proudly unveiled two corned beef sandwiches. Intrigued and delighted, Grissom picked up one of the sandwiches and took a bite. Within moments, small pieces from the crumbly meat started to float around and ran the risk of damaging the delicate instrumentation. After possibly worrying Mission Control with the declaration, 'It's breaking up', Grissom quietly tucked the sandwiches safely away. Back on Earth, some NASA administrators and Congressmen were less than amused with the astronauts' sandwich-based shenanigans and officials eventually declared that: 'We have taken steps . . . to prevent recurrence of corned beef sandwiches in future flights.'[69]

By the mid-1960s, Project Mercury and Project Gemini had run their course, and NASA was ready to move on to a new series of missions that they hoped would culminate with a person on the Moon: Project Apollo. Whereas the Mercury and Gemini missions had involved sending one or two astronauts into space for a relatively short period of time, the Apollo 1 mission would require three astronauts to spend up to two weeks orbiting the Earth. In March 1966, NASA announced that the Apollo 1 crew would consist of Gus Grissom, Ed White and Roger Chaffee.

The Apollo 1 mission called for a capsule that was larger and far more complex than anything that had been built previously. The development of this new capsule (referred to as a Command Module) involved coordinating groups of engineers across the country and proved to be highly challenging. Grissom knew that his life was on the line each time he blasted off into space and he took a hands-on approach to assessing the technology being developed for each mission. When the Apollo 1 capsule started to take shape, Grissom became worried about several flaws, and eventually demonstrated his concerns openly by picking a lemon from a tree in his garden and hanging it outside the Apollo 1 simulator to indicate that he thought the spacecraft was less than perfect.[70]

The mission got the green light, and on 27 January 1967, the three astronauts climbed inside the Apollo 1 Command Module for a routine pre-launch test. The hatch was closed and the simulation began. Within moments, issues began to emerge. Grissom's microphone became stuck in the open position, making it difficult for the various teams to communicate with one another. Annoyed, Grissom remarked: 'How are we going to get to the Moon if we can't talk between two or three buildings?' The countdown was put on hold as technicians attempted to solve the issue.

And that's when the real problems started.

Manfred von Ehrenfried was manning one of the Mission Control consoles during the Apollo 1 test. Born in Dayton, Ohio during the depression, von Ehrenfried joined the Boy Scouts, studied physics, became spooked by Sputnik, heard Wernher von Braun give an inspirational speech on space flight and eventually joined NASA. He went on to play a key role in many of the Mercury and Gemini missions. Fifty years on, the events surrounding the Apollo 1 test are still fresh in his mind:

The Apollo 1 spacecraft was over in Cape Canaveral, and we were in Houston. I was acting as a Guidance Officer, and all through the afternoon the communications were very poor, so we were always sitting there with our hand over our headphones trying to hear the flight test conductor and the crews. We were sending commands to the spacecraft and checking our data. Suddenly I hear 'fire in the spacecraft'. I turned to my fellow Guidance Officer and said, 'Did you hear that?' We all sat there trying to hear what was going on. Then I heard the pad crew's heroic efforts to get the astronauts out.[71]

A few moments into a seemingly routine break, a massive fire had torn through the cabin's pure oxygen atmosphere. Chaffee

98

immediately raised the alarm and the three astronauts tried to escape. The loss of Grissom's previous capsule, the Liberty Bell 7, had resulted in engineers designing an entirely new type of hatch. The Apollo 1's hatch consisted of three separate parts: an inner panel that opened into the craft, a middle panel that formed part of the heat shield, and an outer panel designed to protect the craft during launch. Emergency guidelines suggested that the crew should have been able to quickly remove all three panels and escape from the craft. In reality, this did not prove possible.

Panicked ground crew ran to help the struggling astronauts, but their efforts were hampered by both the intense heat of the fire, and the thick smoke billowing from the spacecraft. Ignoring the obvious danger, the crew bravely battled on and managed to open the hatch within a few minutes. Unfortunately, it was too late. By the time the crews entered the craft, all three astronauts had died from asphyxiation. It took an hour and a half for the bodies to be removed and medics treated many of the rescuers for smoke inhalation.

Von Ehrenfried remembers Chris Kraft looking completely pale and stern. He also recalls the terrible impact that the events had on the mission controllers:

Over in Mission Control we could hear the pad people trying to extinguish the fire. There was nothing we could do. When the noise had calmed down, and we realized what had happened, we all just sat there trying to let it sink in. By then, other staff had heard what had happened, and came pouring in to see if they could help . . . but there was nothing anyone could do. Some of those in Mission Control were never the same. I came out with tears in my eyes, and suddenly realized that this was a serious business.[72]

Ed White was buried with full military honours at West Point Cemetery, New York. Gus Grissom and Roger Chaffee are buried, side by side, in Arlington National Cemetery. A short distance away from their graves, an eternal flame burns to commemorate the death of the grand visionary of America's space programme, President John F. Kennedy.

Although Grissom often appeared to be a man of few words, he kept a detailed journal describing his thoughts and feelings. In one especially poignant paragraph, he reflected on the value of working for the greater good of doing something, no matter how difficult or dangerous that work might be:

'If we die, we want people to accept it. We're in a risky business, and we hope that if anything happens to us it will not delay the program. The conquest of space is worth the risk of life.'[73]

THE KRANZ DICTUM:
A LESSON IN FACING FAILURE

Following the tragic fire, investigators dismantled every part of the Apollo 1 capsule to establish what had gone so terribly wrong.[74] The results were terrifying. The capsule contained a large number of bare and frayed wires. Any of the wires could have caused a deadly spark, which would have then ignited the flammable material inside the capsule, including the astronaut's nylon space-suits, the velcro used to secure tools and even the seats themselves.

Filled with 100% pure oxygen at a high pressure, the flames would have spread extremely rapidly, giving the astronauts little time to escape. Worse still, the complex hatch system made a fast exit impossible that day.

Over the course of the next 18 months, NASA spent millions of dollars making a vast number of changes to the Apollo spacecraft, including designing a hatch that could open in seconds, reducing the amount of combustible materials in the craft and changing the cabin atmosphere from pure oxygen to a hydrogen–oxygen mix. However, the fire didn't just lead to changes in technology. It also caused a fundamental shift in the mindset of the entire organization.

Gene Kranz was perhaps the most recognizable member of Mission Control, and was known for his trademark close-cut flat-top hairstyle. In his fascinating autobiography, *Failure Is Not An Option*, he describes how he called his colleagues together on the Monday morning following the fire.[75] Kranz could see that many were understandably shocked, stunned and upset by the catastrophic incident. He believed that the fire was, in part, the result of people not wanting to openly discuss possible problems with the mission and that this had resulted in the work pushing on when it should have pulled up. Keen to change the organizational culture, he decided to give a brutally honest assessment of the situation. Pulling no punches, and speaking without notes, he stood up and delivered what has come to be known as the 'Kranz Dictum'.

Kranz began by admitting that major problems had emerged during the Apollo programme. He believed that many people were too gung-ho about meeting deadlines and that they ignored these obvious failings. Moving forward, Kranz declared that Mission Control would come to be known by two words: Tough and Competent. By 'tough', Kranz meant that people would feel a strong sense of responsibility for their actions, and would be fully accountable for what they did and failed to do. 'Competent' meant that they would never fall short in terms of their knowledge and skills and never stop learning. Kranz urged everyone to write the words 'tough' and 'competent' on their blackboards, ending by saying that the words would form both a constant reminder of the sacrifice paid by the three astronauts and help ensure that the same tragic situation never happened again.

The Apollo 1 fire helped everyone realize the importance of being open about their worries and concerns. Out went the notion of trying to cover up failure and ignore potential problems. In came the idea of embracing errors and treating them as an opportunity to learn and grow. One flight controller later described the situation as like a poker game where you are playing with your cards face up – you couldn't bluff anyone and instead had to admit failings and mistakes.[76]

The transformation was remarkable – years later many commentators remarked that Neil Armstrong would never have set foot on the Moon had it not been for the hard lessons learned after the tragic Apollo 1 fire.

Jerry Bostick provided a concrete example of this approach in action. During one training session, Bostick and two colleagues made some miscalculations and ruined a mission. During the debriefing session Chris Kraft asked what had happened. Bostick's two colleagues tried to talk their way out of the situation, whereas he simply admitted he had made an error. Later that day Kraft removed Bostick's colleagues from the programme and retained the one person who had been big enough to admit messing up.[77]

In his later years, Kraft also reflected on the important role that learning lessons from mistakes played in the Apollo culture:

There was a tremendous feeling of openness among our organizations. We grew up telling each other we were making mistakes when we made them. And that is how we learned . . . We were never embarrassed about being made a fool of when we made mistakes because we made them. I mean, we made hundreds of them. But we were used to being open about them. And that was fundamental to getting our job done.[78]

Today, the Kennedy Space Center's Visitor Complex contains a wonderful series of exhibits demonstrating the breathtaking technology involved in space exploration. In January 2017, the Center opened a new exhibition to honour Gus Grissom, Ed White and Roger Chaffee, and it contains various objects associated with the three brave astronauts, including models of their spacecraft, their clothing and the tools that they used in space.

At the centre of the exhibition sit the three metal panels that made up the Apollo 1 hatch. The panels have remained in storage for over half a century and, on the face of it, appear to be three innocuous pieces of metal. In reality, they were witness to the darkest moment in the Apollo programme. But the panels also stand as a monument to progress. The Apollo team discovered the importance of being open about their failures and, as a result, learned to grow and move forward. The Apollo 1 fire was a terrible tragedy, but without it, and the change in attitude that it created, Kennedy's vision of putting a person on the Moon before the end of the decade might never have become a reality.

The title of the exhibition perfectly sums up the central message from the tragic fire: '*Ad Astra Per Aspera*' – 'A Rough Road Leads to the Stars'.

*

Learning From Failure

'*The only man who never makes a mistake is the man who never does anything.*'

– Theodore Roosevelt

It isn't possible to turn back the hands of time and discover the precise errors that led to the tragic Apollo 1 fire. Maybe the engineers working on the project had a great deal to do in a very short space of time and rushed ahead recklessly (a situation that some referred to as 'go fever'). Maybe they were scared that if they admitted errors they would look foolish, appear overly negative, not be promoted or maybe even lose their jobs. Maybe the success of the Project Mercury and Project Gemini missions had made them overconfident and gung-ho. However, what we do know is why people are reluctant to embrace their failings and errors in everyday life.

You frequently make assumptions about yourself and others. You might, for instance, have struggled with mathematics when you were young and now assume that you don't have a head for numbers. Or you might meet someone at a party, discover that they work in a library and assume that they are introverted. Around thirty years ago, Stanford psychologist Carol Dweck discovered that when it comes to success, there is one assumption that is vitally important.[79]

Some people assume that their intelligence, personality and abilities are pretty much set in stone. In the same way that a marble statue doesn't change over time, they assume that they are the same person day after day, month after month and year after year. Dweck refers to this as a 'fixed' mindset. In contrast, others assume that their skills and personality are far more malleable and, in the same way that clay is soft and pliable, believe that they will change over time on the basis of effort and experience. Dweck calls this a 'growth' mindset.

Dweck discovered that having a fixed or growth mindset really matters. In one study, she spent several years monitoring hundreds of high school students taking a challenging mathematics course.[80] Although the children with fixed and growth mindsets started out achieving the same grades in initial exams, within just a few months those with a growth mindset were outperforming their fixed friends. The two groups continued to diverge over the next few years, leading to the growth-oriented children eventually obtaining significantly higher grades in their final exams.

Curious, Dweck drilled down into the data and discovered that much of the difference was due to the way in which the children perceived failure. The children with a fixed mindset were convinced that their level of intelligence didn't really change over time and so they were allergic to making mistakes because they thought that there was little they could do to rectify their failings. Eager to appear smart, they tended to avoid more challenging mathematical problems and often hid their failings or blamed them on others. In contrast, those with a growth mindset thought that they could change and had a passion for discovery rather than a hunger for approval. They were happy to tackle increasingly difficult mathematical problems, learn from their mistakes and be more open about their failings.

Dweck soon discovered that the same two mindsets affect many aspects of everyday life. When it comes to losing weight, for instance, people with a fixed mindset believe that they can't change and thus tend to give up the moment they find their hand back in the cookie jar.[81] In contrast, dieters with growth mindsets see such setbacks as a temporary blip and as an opportunity to learn how to avoid temptation in the future. These differing perspectives have a dramatic impact over the long haul, with research showing that, on average, those with a growth mindset lose far more weight than those with a fixed mindset. Similarly, when it comes to romance, those with a fixed mindset believe that people don't really change and so are quick to end relationships when problems emerge. In contrast, those with a growth mindset assume that people can develop and are more prepared to try to work through any issues that arise.

Mindsets also matter when it comes to the workplace.[82] Employees with fixed mindsets tend to see errors as indicative of a deep-seated and unchangeable incompetence and so are much more likely to avoid challenging tasks and cover up their mistakes or blame them on others. In contrast, employees with a growth mindset see mistakes as essential to learning and growth, are happy to venture out of their comfort zones and are more open about their failures.

The same applies to hiring and firing. Leaders and managers with fixed mindsets believe that their employees aren't capable of change and prefer to bring in new talent rather than coach existing employees. They tend to favour jettisoning poorly performing staff, rather than helping them to improve and grow. All of this takes its toll on an organization, creating a culture in which employees avoid situations in which they might perform poorly. In contrast, leaders with more flexible mindsets believe that employees have the potential to develop; they don't expect staff to arrive fully formed, but rather ready to learn. As a result, they are more likely to favour growing talent internally and helping people to do their jobs well.

Of course this doesn't mean that repeated failure is a good thing. What's vital is that people welcome challenging situations and that they are open about their mistakes and learn from their errors. Fortunately, creating a growth mindset isn't rocket science. In fact, it's just a case of taking the two-part astronaut challenge, writing a letter to your imaginary best friend and using some magic words.

The Astronaut Challenge
Part One

In a moment you will be given the opportunity to try to solve a puzzle entitled the Astronaut Challenge.[83] The puzzle is far from easy and around 50% of people fail to solve it in the time available. You don't have to try to solve the puzzle. The choice is yours. Please jot down your thoughts about tackling the tricky puzzle.

My thoughts are: _____

Did you have these sorts of thoughts in your head?:

'Well, that sounds like a difficult puzzle, and I don't want to end up failing, so it's probably best to skip the challenge. Better that than suffer the indignity and stress of possible failure.'

'Great. I love puzzles and this sounds like a good opportunity to demonstrate my genius.'

'I am happy to give it a go. After all, if I fail then no one will ever know.'

As you might have guessed, these types of thoughts reflect a fixed mindset. The good news is that it's easy to develop a much more positive and growth-oriented way of viewing tricky challenges. When you are presented with a tricky challenge, you might be tempted to walk away and avoid the possibility of failure. Instead, spend a few moments thinking about why you should tackle the challenging task. First, what are the benefits associated with success? Perhaps more importantly, what are the upsides of failure? Would you develop some new skills? Or learn something interesting or important? Will taking on the challenge help you to connect with other people? Or might it lead to new opportunities and openings? And if you don't try then you will certainly fail, and where's the dignity in that?

When it comes to the Astronaut Challenge, maybe you will succeed and feel good about yourself. However, even if you fail, you might end up with a fun puzzle that you can share with your friends, colleagues and family. Or the experience will help you solve other challenges in the future. Or maybe you will be presented with the puzzle when you apply for a job and you will be able to score top marks. Or, depending on the type of puzzle it turns out to be, you will learn about a new mathematical principle or psychological technique. Or you will go to a party, discuss the puzzle with a complete stranger, fall in love and end up happily married. All of these sorts of thoughts are associated with a growth mindset.

Assuming that you are now open to taking on the Astronaut Challenge, here goes. Here are three astronauts and three tanks containing electricity, water and oxygen.

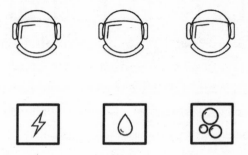

Each astronaut needs all three substances to survive. Can you draw lines that connect the three tanks to the three astronauts' helmets, but in such a way that the lines don't cross one another? So, lines like this would be fine . . .

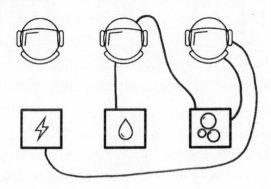

because none of the lines cross one another. However, a line connecting the water to the third astronaut like this . . .

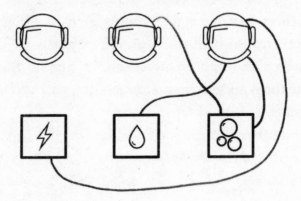

wouldn't be acceptable because it crosses the line connecting the water to the second astronaut. You have precisely 3 minutes to solve the puzzle. Go.

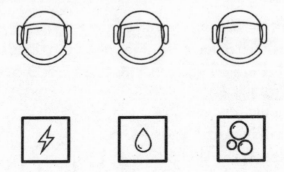

In the next section we'll find out what your performance says about you.

The Astronaut Challenge
Part Two

If you failed the Astronaut Challenge then please don't feel too bad. In fact, everyone fails to solve the puzzle because it's impossible! However, you haven't just wasted 3 minutes of your life, because the thoughts that went through your head after you failed are fascinating.

Did you think one of these sorts of thoughts?

'I am simply not cut out for these sorts of puzzles and that's the last time I try something like that.'

'It's just a silly puzzle, I am just going to ignore it and move on.'

'3 minutes isn't very long and I would have solved the puzzle if I had been given more time.'

These sorts of reactions are an attempt to play down your failure and avoid additional failings in the future. As such, they reflect a fixed mindset. Again, it is easy to shift to a more positive perspective. When people with a growth mindset fail to solve the astronaut puzzle they are eager to discover the solution, and explore how they might be able to use this new-found knowledge in the future. Then, when they discover that the puzzle is impossible, they usually want to know why that's the case (if you are curious, take a look in the Appendix) and think about how they can use the puzzle to annoy their friends and family.

Here is the final part of the exercise. Imagine that you are going to present the Astronaut Challenge to your friends and colleagues – please briefly jot down what you would say to them.

I would say something like: _____

Did you mention that you had failed to solve the puzzle? Remember that people with a fixed mindset are keen to bury their mistakes, whereas those with a growth mindset are more open about their failings. The same problem applies to criticism. When people with a fixed mindset encounter critical feedback, they tend to be defensive, argue back, or attack the person offering advice. In contrast, those with a growth mindset are far more open to change, and so listen in the hope of learning something useful. If you need to switch to a growth mindset, be open about your failings and welcome opportunities to develop.

The same applies within organizations. The mission controllers used to regularly hold lively debriefing sessions at a local German beer garden, and spend hours discussing their mistakes and possible solutions.[84] If you are part of a team, develop a culture in which people are happy to be open about their mistakes and everyone can work together to figure out the best way forward. This type of initiative can take the form of a 'failure' party, an internal website, or flipchart. Perhaps the most creative example of this comes from the social media company Nixon-McInnes. Each month, the company holds a meeting entitled the Church of Fail, where employees are encouraged to 'confess' their mistakes. During each confession, people outline the failure and, more importantly, reflect on what they intend to do differently next time. After each confession, everyone cheers and applauds!

Moonshot Memo
Subject: Damned fool things I have done

Dale Carnegie's classic self-help book, *How to Win Friends and Influence People*, has sold over 30 million copies. Throughout his life, Carnegie kept a personal file entitled 'Damned Fool Things I Have Done', in which he noted down all the important mistakes that he had made and what he had learned from them.[85]

If that technique was good enough for Carnegie, it is good enough for you. Instead of trying to ignore your mistakes, think like the world's greatest self-help expert and briefly jot down each of your errors and what you discovered from making the mistake.

Obviously, it's not great if you keep making the same errors time and again. That's just incompetence. However, it's fine to make mistakes as long as lessons are learned.

Dear Matilda

Imagine that your best friend is called Matilda.[86] You have known Matilda for many years and the two of you have supported one another through thick and thin. Matilda has always wanted to work on space exploration, and a few months ago started a course in rocket science at University. Today you received this email from Matilda:

Hi there,

I hope you are well. I am still enjoying university life, but I am starting to struggle with a few of the courses. Last week I failed an important algebra exam, although I haven't told any of my friends yet. No one said that it would be this hard, and I suspect that my brain simply can't learn some of the mathematical material that's core to the course. I would still love to become a rocket scientist, but I am thinking of dropping out of university.

Any advice?

Lots of love

Matilda

As you might have spotted, Matilda appears to have a fixed mindset. How would you reply, and encourage her to develop a more growth-oriented mindset?

Dear Matilda,

_____

_____

_____

Now let's take a look at your reply. Here are some ideas that you could have mentioned. Did you?

– You could have reflected on how Matilda has changed over the years and how this reflects an amazing ability to grow and learn. Perhaps she gained a new physical skill, such as learning to drive, playing the guitar, juggling or play golf. Or maybe she acquired a new mental ability, such as learning to speak a new language, gaining a qualification or adapting to a new role at work. How could you use these experiences to make Matilda realize that she can change and grow? How would this help her to keep going?

– Did you reflect on what caused these changes to happen?

Maybe Matilda spent a long time practising, working hard and being resilient? Did she face failure and learn from mistakes? Did she manage to keep on going, even when the going got tough? Did she ever find it helpful to admit her failings and use them to identify and improve weaknesses? Again, how would this perspective help her to continue with the course?

– What about your own life? What physical skills and mental abilities have you managed to master over the years? How have you changed over the past decade or so? Once again, how did these changes happen? And how could that motivate Matilda?

– Given that Matilda thinks that she doesn't have the right type of brain for mathematics, it might have been helpful to mention the scientific evidence showing that all brains have an amazing ability to change (a phenomenon known as 'neuroplasticity'). For instance, if you are unfortunate enough to have a stroke, your brain will reorganize itself to help recover some of the abilities that you have lost. If you wear a blindfold for a few days, the part of your brain that processes visual information starts to focus more on sound and touch. If you spend a couple of years mastering a musical instrument you will boost the networks of neurons responsible for attention and decision-making. And spend just a couple of weeks learning to juggle and your brain grows more nerve cells in the area involved in linking sight to movement.

– Finally, you could have included an inspirational illustration from history. Look back at the previous chapter for several striking instances of people who, for instance, ran faster than anyone thought possible, promoted women's rights despite being blind and deaf, or were born into poverty but became one of the richest people in the world. All of them demonstrated the importance of persevering and the potential for change.

OK, now we have reviewed these key points, write a second reply to Matilda incorporating some of these ideas, and encouraging her to adopt a growth mindset and to treat her failure as a springboard to success.

Dear Matilda,

_____

_____

_____

Magic Words

Magicians use magic words like 'Abracadabra' and 'Hocus Pocus' to apparently achieve the impossible. Similarly, you can use certain words and phrases to magically promote a growth mindset when you talk to yourself and others.

Imagine that you have a son or daughter and that they have just taken an important exam and obtained a good grade. What would you say to them?

I would say something like: _____

Now let's imagine that they hadn't really worked very hard but nevertheless obtained a good grade. What would you say then?

I would say something like: _____

Finally, what if they had worked very hard and didn't do well? What would you say to them?

I would say something like: _____

In the late 1990s, Carol Dweck carried out a remarkable study into mindsets.[87] She asked a group of students to work on some mathematical problems and then split the students into two groups. The students in one group were praised for being intelligent ('You must be pretty smart to do so well . . .'), while the others were praised for their effort ('My, you seemed to have worked hard on this . . .'). Remarkably, this small change was enough to change the students' mindsets, with those praised for their intelligence being more likely to develop a fixed mindset, and those praised for their hard work developing a growth mindset. Next, all of the students were given another exam and asked how they had done. Almost 40% of the students who had been given the gift of a fixed mindset tried to cover up their failings and knowingly inflated their scores. Among the students with a growth mindset, this figure fell to just 10%.

Think back to what you said to your son or daughter when they obtained a good grade. Did you praise their intelligence, talent and ability? As we have discovered, praising people for being naturally talented certainly makes them feel good, but they then start to worry about tarnishing their newfound reputation and so become more likely to hide their errors and cover up their mistakes. Instead, it's much better to praise effort by focusing on their learning skills ('Your idea of testing yourself every night really paid off'), or exam technique ('I was impressed with how you learned to perform under pressure').

117

What did you write when your son or daughter hadn't worked very hard but nevertheless obtained a good grade? Once again, it isn't a case of praising their ability, but rather urging them to up their game ('OK, it looks like that was too easy for you. What would challenge you?').

If they have worked hard and didn't do well, the most productive approach is to focus on their effort and moving forward ('I really liked the fact that you put so much effort in – shall we try to figure out what you struggled with?').

The same idea applies in the workplace. To encourage those around you to adopt a growth mindset, try to give positive feedback that emphasizes hard work ('Congratulations, you put in the effort to make the deal happen'), willingness to take on new challenges ('Moving out of your comfort zone really worked well') and resilience ('That was tough, but you kept bouncing back and got there in the end. Well done').

Finally, when it comes to talking to yourself and others, Carol Dweck urges people to embrace the magic word 'yet'.[88] If you find yourself thinking that you can't achieve a goal, or hear someone else say that they are struggling to be successful, help to promote a growth mindset by adding the word 'yet' to the end of a sentence. For instance, 'I am not the sort of person that goes to the gym' becomes 'I am not the sort of person that goes to the gym, **yet**'. Similarly, 'I don't have enough money to start my own business' becomes 'I don't have enough money to start my own business, **yet**'. And 'I don't have a University degree' becomes 'I don't have a University degree, **yet**'.

When it comes to creating mindsets, the way in which you speak to yourself and others really matters.

SUMMARY

Your attitude towards failure matters and it's important to adopt a growth mindset, accept tricky challenges, and be open about errors and mistakes.

– See difficult challenges as an opportunity to develop and learn. Remember that your comfort zone might feel like a pleasant place to live, but nothing grows there.

– When you fail, be honest with yourself and others. Don't try to cover up errors or pretend that you did well. Find out what went wrong, and ensure that you don't make the same mistake in the future. Similarly, when it comes to constructive criticism, avoid arguing back, and instead remind yourself that there's much to be gained from listening and learning.

– Try following in Dale Carnegie's footsteps and keeping a list of all the damned foolish things that you have done and what you learned from them.

– Remember that certain types of language help promote a growth mindset. When giving praise, stress effort over achievement and learning over talent. And don't forget to use the magic word 'yet'.

Moonshot Memo
Subject: Do good work

Although Grissom was more than able to handle Project Mercury's punishing training programme, he struggled with some other aspects of being a world-famous astronaut.

As a fighter pilot, Grissom had learned to convey his thoughts using the fewest words possible.[89] While this was fine when he was flying over enemy territory, it sometimes proved to be an issue when he was chatting to friends and colleagues on the ground.

On one occasion, the Mercury Seven astronauts were visiting a rocket manufacturer and Grissom was asked to say a few words to a large audience of engineers. Never a natural public speaker, he walked up to the podium and uttered just a few words: 'Well . . . do good work.' Fortunately, the engineers loved Grissom's simple motto and it quickly spread throughout the aerospace industry, giving rise to 'Do Good Work' signs and posters in factories and offices across the nation.

Grissom's quip shows the power of the motto. What simple phrase sums up your approach or project?

5

'IT WON'T FAIL BECAUSE OF ME'

*Where we come across the mantra that ensured the
world's largest rocket got off the ground, and uncover
how you can develop the attitude that gives you altitude.*

The tragic Apollo 1 fire was a major setback. However, important
lessons were learned and eventually the Apollo programme started
to get back on track. Project Mercury had put one astronaut in
orbit around the Earth and Project Gemini had added a second
crew member into the mix. To be successful, the Apollo flights
would have to involve a three-person crew travelling thousands of
miles across space, landing on the Moon and returning safely to
the Earth. To make this giant leap, Team Apollo would depend on
a series of technological breakthroughs and an equally remarkable
mental attitude.

When it comes to blasting rockets into space, Cape Canaveral
has been the American launch pad of choice since the end of
the Second World War. The Cape is located in Florida's Brevard
County and many of the local residents are passionate about space

exploration. In the late 1990s, space fan and long-time Brevard County resident 'Ozzie' Osband found out that there was going to be a public hearing about new telephone codes in the area and came up with a brilliant idea. Osband took the day off work, went along to the hearing and passionately argued that Brevard County be assigned the code '3–2–1' to celebrate the rockets that had regularly blasted off from the Cape. Good sense won the day and every long-distance telephone call to Brevard County now reminds callers of the area's important contribution to the exploration of space. Appropriately, Osband was later assigned a telephone number that reflects his passion and dedication to rocketry: '321-Liftoff'.[90]

Throughout the 1960s, people travelled to Florida's space coast, excitedly waited for announcers to read out Osband's telephone number, and then experienced the thrill of a live rocket launch. The launches that took place during Project Mercury and Project Gemini were exhilarating and uplifting. However, the rockets involved in these missions paled in comparison to those required by the Apollo programme. At the end of Project Gemini, two-person capsules were being sent into orbit around the Earth. In contrast, the Apollo missions would involve a Command Module capable of transporting three astronauts to the Moon, a Lunar Module that would ferry them to and from the lunar surface and a Service Module that would contain their supplies and fuel. To ensure that this amount of weight could escape the Earth's gravitational pull, the Apollo engineers had to design and build the world's tallest, heaviest and most powerful rocket: the Saturn V.

Every statistic associated with the Saturn V is staggering. The rocket consisted of five main sections. The bottom section was an enormous 42 metres tall and 10 metres in diameter. It housed over 300,000 gallons of liquid nitrogen, over 200,000 gallons of kerosene fuel and 5 of the largest engines ever constructed. The second section stood 25 metres high, contained a further 260,000 gallons of liquid hydrogen, 80,000 gallons of liquid oxygen and another five massive engines. The third section stood 18 metres tall and contained a further 65,000 gallons of liquid hydrogen and 20,000 gallons of liquid oxygen. Then came the penultimate section, which housed the Command Module, the Lunar Module and the Service Module. Finally, topping off the Saturn V was an escape rocket that was designed to pull the astronauts free should there be a malfunction during lift-off.

The Mercury and Gemini missions had blasted away from relatively modest-sized launch pads at Cape Canaveral's Air Force Station. However, the Saturn V rockets required a much larger launch pad and infrastructure, and NASA decided to construct an entirely new facility on a peninsula just north of the Cape. Construction began in November 1962 and was completed in late 1967. In 1963, just a few weeks after President Kennedy's tragic assassination, the decision was made to rename the facility the Kennedy Space Center in honour of Apollo's visionary architect.

Each of the Saturn V sections were constructed on sites across America and transported to the Kennedy Space Center by road, barge and air. Once there, the sections were joined together in the astonishingly big Vehicle Assembly Building. One of the largest structures in the world, the outside of the building is home to the biggest painting of an American flag (each star is six feet across and each stripe is nine feet wide), and the inside is so vast that it produces its own weather (on humid days, rain clouds appear under the ceiling, and have to be removed by the building's 10,000 tons of air conditioning equipment).

Once assembled, each magnificent Saturn V rocket was 35 storeys high, twice the height of the rockets that would later launch the Space Shuttle into low Earth orbit. After engineers had carried out a complex series of checks, the gigantic rocket and its supporting tower were slowly transported to the launch pad. On launch day, the astronauts were strapped into their spacecraft and the ground crew retreated to a safe distance.

The Saturn V consisted of more than six million parts and, if only a tiny percentage of them failed during the launch, the enormous amount of fuel in the rocket ran the risk of exploding and turning the entire launch pad into a huge fireball. Indeed, in 1969, a similar-sized Russian rocket exploded during blast off, and resulted in one of the largest non-nuclear explosions in history. The Apollo astronauts were perched on top of what was essentially a gigantic bomb and even the smallest of mistakes could prove fatal.

ON BEING ACCOUNTABLE FOR WHAT YOU DO AND WHAT YOU DON'T DO

Before one of the later Apollo launches, astronaut Ken Mattingly spent a few nights going out to the launch pad and studying different parts of the rocket that he hoped would take him to the Moon. He often thought about the thousands of people involved in the design, fabrication and checking of each part of the Saturn V.

One night, Mattingly took the lift to an upper level of the launch tower and found himself outside an open hatch.[91] There, he climbed through and entered a large room packed with pipes, cables and wiring. A lone technician in the room recognized Mattingly and the two of them started to chat about the risks involved in the mission. During the conversation the technician explained that he had no idea about how many parts of the rocket worked. For instance, he didn't understand how the enormous amount of fuel created the force required to escape the Earth's gravitational pull or how the navigational systems would guide it to the Moon. Understandably, Mattingly became somewhat concerned. However, the engineer then continued and explained that the panel in front of him *was* his responsibility. It was his job to fully understand the complex electronics inside that panel and ensure that they were in perfect working order. The engineer ended by assuring Mattingly that when it came to that panel, the project wouldn't fail because of him.

It was in that moment that Mattingly realized that the Apollo missions had been successful because so many of those involved in the project had exactly the same sense of personal responsibility. Pad leader Günter Wendt is another striking example of this 'it won't fail because of me' attitude.

Thin, bespectacled, German-born and with a fondness for bow ties, Wendt was one of Apollo's best-loved and most eccentric characters. Wendt was in charge of the Saturn V White Room, the small area used by astronauts to make their final preparations before entering the spacecraft.

It was Wendt's job to ensure that the astronauts were safely buckled into their spacecraft, say a last goodbye and good luck and then seal the Command Module's hatch. Wendt ruled the White

Room with a pleasant smile and an iron fist. Nobody touched any-thing in the area without his permission and when one engineer went to make a change without Wendt's consent, he called security and had the engineer removed.

The Apollo team held him in warm regard and frequently joked about his strong sense of responsibility, with astronaut John Glenn assigning him the nickname 'der Führer of der Launch Pad'. In the same way that the technician who chatted to Ken Mattingly felt a strong sense of responsibility in his job, so Wendt was not one to pass the buck. Before John Glenn's flight, Wendt told Glenn's wife, Annie:

> I cannot guarantee the safe return of John. Nobody can. There's too much machinery involved. The one thing I can guarantee you is that when the spacecraft leaves, it's in the best possible condition for a launch. If anything should happen to the spacecraft, I would like to be able to come and tell you about the accident and look you straight in the eye and say, 'We did the best we could.' My conscience then is clear.[92]

Over in Houston, many of the mission controllers had exactly the same approach to their work. When Neil Armstrong set foot on the Moon, Ed Fendell was in charge of the communications between ground control and the astronauts. Like so many of those involved in Mission Control, Fendell was from a modest back-ground. Born and raised in Connecticut, he spent much of his childhood working in his father's grocery store and his family never had a great deal of money. After obtaining an Associate Degree in Merchandising and working in Air Traffic Control,

Fendell eventually ended up at NASA. Fendell was one of many mission controllers to reflect on the importance of taking responsibility, not wanting to mess up, and working hard:

> It would be wrong to think that we got to the Moon because we were especially smart. Many kids today are way brighter than us, and have much better tools. We got to the Moon because we had the right attitude. Everyone in the team had a strong 'can do' attitude. I never heard anyone say, 'I can't do that' – you were expected to find solutions to problems and provide the appropriate work around. We had a strong work ethic. We could be there until 9 o'clock at night, and work Saturdays. It was more than a job, it was a way of life.[93]

This attitude pervaded the whole of Mission Control. If someone said they were going to do something, they did it. They didn't procrastinate, pass the buck, or cut corners. They were as conscientious as they were hard working, and their word was their bond.

Glynn Lunney was part of NASA from the early days and played a key role in the success of the Mercury, Gemini and Apollo missions. He was working in Mission Control during many of the most historic moments of the space race, and is hugely respected across the community. Lunney spoke about the importance of selecting mission controllers on the basis of their 'can do' approach and sense of responsibility:

> We got lots of applicants and they all had decent grades, but I was far more interested in their attitude. I was looking

for people who really wanted to be part of the Apollo mission, and would do anything to make it successful. Often, they weren't perfect students, but they had the right attitude, the perfect attitude. Sometimes, if their grades weren't good enough, I would hire them based on their positive attitude. Everybody was absolutely passionate about getting their job done and getting it done right.[94]

And it wasn't just a case of recruiting the right people. The Apollo managers trusted everyone to do their jobs and this imbued them with a strong sense of responsibility. Lunney saw the power of this approach first hand:

Today we talk about leadership by example, charisma, or fear. This was leadership by respect. The managers made everyone feel trusted, and so people felt a strong sense of loyalty and did the best job that they could do. As I look back, I realize that that's the best equation for leadership. It felt magical, and I saw it in action again and again.[95]

Flight Controller Jerry Bostick expressed the same sentiments when he spoke about those leading Mission Control:

We really didn't want to mess up. Our leaders trusted us, and we didn't want to let them down. You were working for living legends like Chris Kraft, and when he gave you a job to do, you felt that you had his ultimate trust. Kraft would say 'Here's what I want you to do, and I want you to have it done in three weeks, if you need any help, give me a call but otherwise I'll see you in three weeks.' It's ultimate trust. You would walk out of his office thinking, 'I can't let that man down.'[96]

Jay Honeycutt, another Apollo engineer who worked closely with Mission Control, revealed how the strong sense of responsibility trickled down at every level:

I was twenty-seven when I was working on the Apollo program. Chris Kraft took a bunch of people that were around my age, and gave them an incredible amount of responsibility. His message was 'You've got this thing, and you'd better not listen to the Flight Directors, because if they tell you wrong and you do it wrong, then I'm going to blame you, I'm not going to blame them. This is your little spot of responsibility.' I think Kraft was right – you develop people by giving them responsibility and giving it to them early on in their careers.[97]

'I VONDER VERE GÜNTER VENDT?'

The investigations into the tragic Apollo 1 fire resulted in manned flights being suspended for around a year and a half. During the hiatus, the Apollo team launched several unmanned Saturn V rockets and, except for a few shudders and engine failures, all went well. In late 1968, engineers believed that they had solved the legion of issues raised by the fire and gave the green light to Apollo's first manned mission. The mission aimed to give the new Apollo technology a full workout in space and involved spending eleven days in orbit around the Earth. With the Saturn V test launches taking up the earlier numbers, the first manned mission became Apollo 7.

On 11 October 1968, Apollo 7 astronauts Wally Schirra, Donn Eisele and Walter Cunningham climbed into their spacesuits and were driven over to the Saturn V launch tower. They took the lift up to the top of the tower, made their way along a walkway, and entered the White Room. There, under the ever watchful eye of pad leader Günter Wendt, the three astronauts made their final preparations before entering the Command Module. Wendt wasn't working as the pad leader when the tragic Apollo 1 fire happened and, like everyone involved in the programme, was devastated by the deaths of the three astronauts. Astronaut Wally Schirra had insisted that Wendt was given control of the White Room for the Apollo 7 flight.

Once Wendt had ensured that the astronauts were safely inside the Command Module, he shook their hands and closed the hatch. In an attempt to lighten the atmosphere, Eisele adopted a mock German accent and quipped: 'I vonder vere Günter Vendt?'[98]

On 11 October 1968, Schirra, Eisele, and Cunningham felt five of the world's most powerful engines burst into life beneath them. The sound of the engines was deafening and over 900,000 gallons of water (the equivalent of one and a half Olympic swimming pools) per minute had to be pumped under the launch pad to absorb this vast amount of noise and so prevent it ripping apart the rocket.

Within moments the engines produced 7.5 million pounds of thrust and the gigantic Saturn V rocket started to lift away from the Earth. Two minutes later, the rocket had accelerated to around 9,600 kilometres per hour. When the Saturn V reached a height of about 60 kilometres, the lower fuel tanks were empty, and the first section separated away from the rocket and fell into the sea. Thirty seconds later, the five engines on the second stage of the rocket kicked in, accelerating the Saturn V to 24,000 kilometres per hour and up to a height of around 176 kilometres. Nine minutes into the flight, this second stage was out of fuel, separated from the rocket and fell back to the Earth. Finally, the single rocket on the third section ignited, propelling the remainder of the rocket to an astonishing 28,000 kilometres per hour. Around eleven minutes after blast off, Schirra and his crew were now in orbit around the Earth.

Apollo 7 is widely considered to be one of the longest and most ambitious test flights in history. The crew had a packed schedule that involved testing endless operating procedures and carrying out lots of experiments. In addition, they were the first astronauts to transmit a live television broadcast from an American manned spacecraft.

When he was on the ground, astronaut Wally Schirra could be a jovial character. A firm believer in the old adage that 'Levity is the lubricant of crisis', Wally – or, as he was known to some, 'Jolly Wally' – had a reputation for having fun and pulling pranks.[99] Unfortunately, on Apollo 7, Jolly Wally was not so jolly.

The Apollo 7 capsule was designed to protect the crew if they aborted during the launch and landed in the sea. However, it ran the risk of not providing adequate protection if the crew aborted and came down on land. As a result, it was important that the wind conditions were right for the launch. A few hours before lift-off, Schirra became concerned about the direction and strength of the wind, and wondered whether it might be better to delay the launch.[100] After a considerable discussion, the team decided to go ahead and the launch proceeded safely. Nevertheless, the episode understandably made Schirra feel uneasy. Unfortunately, worse was to come.

Schirra developed a severe head cold on the second day of the mission, and soon passed the illness to Eisele. Worse still, the lack of gravity meant that various bodily fluids didn't drain away as they do on the Earth, leaving Schirra and Eisele with blocked ears and stuffy noses. The cold caused Schirra to become short-tempered and that, coupled with some terrible in-flight food and a foul-smelling, waste-collection system, led him to challenge some of the instructions coming from Mission Control.

The situation came to a head as the astronauts prepared for re-entry. Schirra insisted that he wouldn't wear his space helmet during the descent because he was worried about his blocked sinuses bursting his eardrums and so wanted to be able to blow his nose. After a fierce battle with Mission Control (Houston: 'It's your neck, and I hope you don't break it'), the three astronauts safely completed their re-entry sans helmets. Shortly after the flight, Schirra made the best of a bad situation by appearing in several television commercials for a cold medicine, assuring viewers that each capsule provided highly effective relief from sneezing and a blocked nose.

Despite the head colds and tension, the Apollo 7 mission was a spectacular success. The terrifying Saturn V rocket had blasted off exactly as planned and almost all the equipment and procedures had worked like clockwork. The work had been complex and demanding, and even the smallest of slips could have resulted in disaster.

The Apollo programme proved to be a huge success because people were highly conscientious. They had a strong sense of responsibility and their attitude can be summed up in one powerful mantra: 'It won't fail because of me'.

Moonshot Memo
Subject: Jolly Wally

Look into the mission archives and you will soon find evidence of Wally Schirra's sense of fun.

A few days before Christmas 1965, Schirra and his colleague Thomas Stafford blasted off on a Gemini flight. During the flight, Schirra and Stafford informed Mission Control that they had just seen an unidentified flying object:

'We have an object, like a satellite going from North to South, probably in a polar orbit. He's in a very low trajectory travelling from North to South and has a very high climbing ratio . . . Looks like he might be going to re-enter soon.'

As Mission Control frantically searched their screens for evidence of this mysterious sighting, the controllers were astonished to hear 'Jingle Bells' being played through their headsets.

The two astronauts had managed to secretly smuggle a tiny harmonica and some bells on board, and were merrily playing away. The controllers realized that the Santa sighting and jolly tune was a prank, and enjoyed the moment.

Schirra and Stafford were the first people to play musical instruments in space and their harmonica and bells can now be seen in the Smithsonian National Air and Space Museum.

Also, during the mission Schirra had been required to vent his urine out into space. He had noticed that the urine instantly froze and was transformed into a spray of tiny golden globes. Schirra took several photographs of the spray and once he was back on the Earth, attempted to convince astronomers that the images were evidence of a new constellation, 'Urion'.[101]

It's important to have a strong sense of responsibility and to take matters seriously. However, remember to lighten up from time to time. But do keep it fun for everyone, with research showing that negative humour (such as insults or sarcasm) often causes divisions whereas more positive humour is good for morale, optimism, relieving stress, and bonding people together.[102] Jolly Wally's pranks were inclusive and playful – follow in his footsteps and make sure that everyone has a good time.

*

Developing The Attitude
That Gives You Altitude

The psychological roots behind the 'it won't fail because of me' attitude lie in the groundbreaking work of Californian psychologist Lewis Terman.[103]

Born in 1877, Terman was fascinated with the notion of genius and dedicated his career to trying to figure out whether being bright is more biology than upbringing. After completing his doctoral thesis on the topic ('Genius and stupidity: A study of some of the intellectual processes of seven "bright" and seven "stupid" boys'), Terman was offered a position at Stanford University, and began to work on what was to become one of the longest-running studies in the history of psychology.

In the 1920s, Terman scoured schools across California, identified over a thousand highly intelligent children and started to track their lives. Every five years or so, team Terman contacted these volunteers (who came to be known as 'Termites'), and asked them to complete a variety of psychological tests to describe what had been happening in their lives.

Terman's achievements are many. He coined the phrase 'intelligence quotient' (or 'IQ' for short), pioneered an experimental technique that involved tracking people over time (known as a 'longitudinal study'), and inspired an enormous sense of loyalty among his volunteers. (In the early 1940s many of the Termites were fighting in the Second World War and filled out their questionnaires from foxholes on the front line). Over time, Terman and his successors were able to look back on their remarkable data and explore the relationship between intelligence and success. To some, the results seemed compelling evidence that being bright is vital for doing well in life. After all, many of the highly intelligent Termites ended up with good incomes and, in some instances, even made their way into *Who's Who in America*. However, critics were quick to point out that this success might have been due, at least in part, to the psychological boost that the Termites received from being part of the study. As the debate continued, researchers dug deeper into the data and started to examine some of the other factors associated with success. They were in for a surprise.

When psychologists measure personality, they tend to focus on five key traits: Extraversion (sociability), Openness (creativity), Agreeableness (friendliness), Neuroticism (emotionally stability) and Conscientiousness (self-discipline). When the research teams looked at the factors related to success, conscientiousness consistently came top of the class and frequently beat both intelligence and the other aspects of personality into poll position.[104]

Conscientiousness quickly became a hot topic, with researchers across the globe publishing over a hundred scientific articles on the subject. No matter where they looked and who they studied, the same positive pattern of results emerged. Conscientious people tend to achieve better grades in school and college, are less likely to be involved in crime, are more likely to find and retain employment, are especially likely to work their way to the top and enjoy higher salaries, and have especially happy relationships and marriages.[105] In fact, even being around a conscientious person is good for you.

Psychologists from Washington University in St. Louis spent five years studying how a person's income depended on the partner's personality. Once again, conscientiousness mattered most – having a conscientious partner was associated with higher income, greater job satisfaction and a remarkable increase in the likelihood of a promotion.[106]

Like all of the major facets of personality, conscientiousness reflects a range of behaviours, including being punctual, meeting deadlines, working hard, not procrastinating, and having a strong sense of honesty and integrity.[107]

Conscientious people tend to be organized, and so are better able to prepare for exams, interviews and assignments. They also adopt healthy habits and so tend not to smoke, drink irresponsibly or be reckless drivers. They get stuff done, and do it properly, leading to them being offered larger and more important jobs and projects; their word is their bond, resulting in them enjoying the benefits that flow from being trusted and held in high regard. In short, you can be certain that a project will never fail because of them.

The good news is that researchers have discovered that everyone can boost their levels of conscientiousness. It's just a case of getting in touch with your inner control freak, preventing procrastination, adopting the seven habits of truly successful people and remembering an old joke about an inflatable school.

Getting in Touch with Your Inner Control Freak

Please read these pairs of statements and, for each pair, circle the statement that you agree with the most.

Column A	Column B
Many of the negative events in people's lives are due to chance.	People's seemingly bad luck is actually due to mistakes that they have made.
Wars will always be a fact of life, no matter how hard we try to stop them from happening.	If people were more interested in politics, there would be fewer wars.
Fate and destiny often dictate what happens in life.	People are able to guide their own lives through their decisions and actions.
Great leaders are born, not made.	People become successful leaders through experience and hard work.
Exam questions are often unpredictable and so it's not worth studying hard.	If you are a student, and have revised well, then most tests are fair.
When it comes to getting a good job, it's usually a matter of being in the right place at the right time.	People who work hard, and make the most of opportunities, end up with more successful careers.

Around the time of the Apollo programme, psychologist Julian Rotter came up with a concept known as 'locus of control'.[108] According to Rotter, people can be classified in relation to the degree to which they believe that they are in control of their lives. More specifically, Rotter thought that everyone sat somewhere on an 'external' to 'internal' continuum. People with a strong external locus of control tend to blame outside forces for their circumstances (such as chance, powerful governments, the establishment, management or reptilian humanoids), while those with a strong internal locus of control tend to believe that the events in their lives are dictated by their thoughts and actions.

The questionnaire that you just completed is based on Rotter's work. To find out where you sit on the continuum, take at look at the statements that you have just circled. How many of the statements did you choose from Column A and how many from Column B? If you have a strong external locus of control then you will have chosen all six statements in Column A, whereas if you have a strong internal locus of control you will have chosen all six statements from Column B. Most people fall somewhere between these two extremes.

Over the years, thousands of people have completed these sorts of questionnaires in studies and experiments the world over.[109] In general, people with an external locus of control don't believe that their own efforts will change anything, and so frequently feel hopeless and powerless. In contrast, those with an internal locus of control believe that they can shape their lives and are masters of their destinies. Compared to their external counterparts, people with an internal locus of control are happier, healthier and more successful. As a result, they tend to take responsibility for their actions, work hard to achieve what they want, persist in the face of failure, and finish what they start.[110]

The same applies to matters of success and failure. We all have a tendency to believe that our successes are due to our own hard work and inner genius and that our failings are the result of bad luck and other people. However, highly conscientious people are more prepared to take responsibility for the bad times as well as the good, and so are seen as more dependable and trustworthy.[111]

Of course, you can't control everything. Some aspects of your life really are down to chance, your upbringing, or other people. However, when it comes to promoting conscientiousness, it's generally better to get in touch with your inner control freak, and focus on the power that you have to shape your world.

Preventing Procastination

'Tomorrow is often the busiest day of the week.'
– Spanish Proverb

You know the feeling. You have an important task to do at the start of the day, and yet suddenly find yourself: checking your email, watching a video of a skateboarding duck, nipping out to buy a new notepad, making a cup of coffee, taking the dog for a walk, changing the light bulb in the downstairs toilet, listening to a podcast, taking the dog for another walk, making dinner, telephoning a friend for a chat, darning your socks, oiling the creaking door that has been annoying you for ages, taking the dog for the final walk of the day and going to bed.

Surveys show that around 95% of people procrastinate (with the remaining 5% saying that they will answer the survey the following day). But why do you do it and, more importantly, what can you eventually get around to doing about it?

Conscientious people have an amazing ability to avoid procrastinating, and there are several techniques that you can use to stop delaying and start doing.

Think about the future

In an episode of *The Simpsons*, Marge tells her husband Homer that one day their kids will move out and that he will regret not spending more time with them. Homer pours himself a drink and explains that that will be a problem for 'future Homer', and that he really doesn't envy that guy.

Most people tend to think like Homer and enjoy living in the moment, rather than dwelling on the long-term consequences of their actions. They reach for a delicious cream cake rather than think about the long-term weight gain or light another cigarette rather than dwell on the negative impact that smoking is having on their health. This type of 'here and now' thinking also encourages procrastination, with, for instance, procrastinators choosing to head out for a fun evening rather than submit their tax return on time, or play enjoyable computer games rather than work on an important project.[112]

On the upside, research shows that spending a few moments shifting to a more long-term perspective has a surprisingly dramatic effect. A few years ago, psychologist Hal Hershfield recruited a group of young volunteers, asked them to go into a virtual reality laboratory and look into a mirror. Some of the volunteers were encouraged to focus on the present and saw a regular image of themselves in the mirror. In contrast, others were encouraged to think about the future and were presented with a digitally aged image of themselves that made them look as if they were in their twilight years. The volunteers were then given $1,000 and offered an opportunity to invest some of the money into a long-term savings account. Those that had encountered their aged selves placed far more money into the account.[113]

The same applies to procrastination. To avoid procrastinating you might, for instance, think about how failing to start work on a project will cause extra stress as the deadline approaches. Or how putting off your diet will result in weight gain and possible illness. Or how failing to answer that backlog of emails will decrease your chances of a promotion. Or how not starting to write that report will mean that other people will have to work harder and so will end up thinking badly of you.

Think about the problems that will emerge in the future
if you don't make a start right now.

Take it one brick at a time

Apollo Flight Controller Jerry Bostick once shared a simple, but powerful, insight into productivity: 'Don't do nothing, just because you don't have time to do everything you want to do.'[114] The same applies to procrastination; people often become overwhelmed with the size of the task in front of them and so ending up feeling paralysed and doing nothing.

To overcome the problem, do what builders do – take things one brick at a time. Losing five kilograms in weight might seem an impossible aim and might encourage you to procrastinate. However, set a goal of just half a kilogram a month, and away you go. Similarly, launching a successful startup may have a demotivating effect, whereas spending one evening a week working on your new venture seems far more manageable. The important thing is to make a start, because then it will be much easier to keep on going. Also, extending the building analogy, see yourself like a hard-working builder who is willing to work in all weathers. Come rain or shine, whether you feel like it or not, you have to lay your foundations, mix the cement, and start to build walls. Avoid believing that you have to be in just the right mood or state of mind to get started. Instead, ignore how you feel and just get on with it.

Be like a good builder – take it one brick at a time
and work in all weathers.

Create smart deadlines

There's more to deadline setting than meets the eye. A few years ago, Professor Yanping Tu from the University of Florida discovered that people tend to classify deadlines in terms of whether they are days, weeks, months or years away. In one study, volunteers were offered a reward if they opened a bank account within the next six months. Some of the volunteers were presented with the task in June, with a December deadline. Others were asked in July and given a January deadline. Far more of those in the 'June to December' group completed the task. Why? Because when the task spilt over in the next calendar year, the volunteers perceive it as something that can be put off until a later date.[115] Similarly, in another study, one group of volunteers were given a task on a Monday, and told that it had to be completed by Friday, whereas another group were given the same task on a Thursday and told that it needed to be done by the following Monday. Once again, people procrastinated more when the deadline was the following week.

Try to create deadlines that don't spill over into the following week, month or year. In addition, change how you think about those deadlines by counting the number of days you have to do the task. Some of Professor Tu's other work suggests that it is helpful to mark the deadline on a calendar, colouring all of the days between then and now in the same colour and then numbering them.

Finally, make deadlines as specific as possible. Don't tell someone that you will have that report with them by the end of the week. Instead, let them know that it will be on their desk by 3 p.m. on Friday.

Create deadlines that feel close and pressing.

Moonshot Memo
Subject: Procrastination

The mission controllers isolated themselves from the outside world and became completely focused on getting to the Moon. How about following in their footsteps by minimizing some of the distractions around you. Switch off your email, telephone and social media and avoid working anywhere near a television or bed.

The 7 Habits of Truly Successful People

Over the years, psychologists have identified the habits of highly conscientious people. Incorporate them into your life and see what happens.

Habit 1: A place for everything . . .

Conscientious people are well organized. At home, they make their beds in the morning, wash their dishes directly after a meal, and take out the rubbish before it overflows. At work, they keep their desks tidy and their paperwork filed away. In addition, they plan their day, and help remember important information and dates by making lists and regularly consulting their calendar.

– Be organized. In your workplace, file away any piles of papers, put pens and pencils in a drawer marked 'pens and pencils', and get rid of those half-full coffee mugs. Take five minutes each morning to plan the day ahead and clear your desk when you have finished working each evening. Also, get into the habit of writing down important information. Entrepreneur Richard Branson always carries a notebook with him, Oprah Winfrey has kept a handwritten journal for most of her life, and George Lucas regularly jots down his ideas in a notebook.

Habit 2: Half the time

In 1955, British historian and author Cyril Northcote Parkinson suggested that work expands to fill the time available for its completion. This idea, which has come to be known as Parkinson's Law, has been

144

put to the test and come up trumps.[116] Conscientious people understand that shorter deadlines can encourage them to find innovative ways of streamlining tasks and stop them wasting time.

– *Decide how long a task should take, and then give yourself half that time. Scheduled an hour-long call? Cut it to 30 minutes. Have a writing assignment pencilled in for three days? Aim to do it in two days. Even if you don't get everything done, powering through the majority of the task will leave time to focus your attention on the more problematic elements.*

Habit 3: Arrive 10 minutes early

Conscientious people are punctual people.[117] They don't miss meetings, cancel at the last minute or arrive late. Part of the reason is that they tend to wear a watch and watch wearers are especially likely to be on time.[118] In addition, they have a more accurate perception of how long they have to get somewhere. A few years ago, San Diego State University psychologist Jeff Conte examined the way in which two groups of volunteers perceived time.[119] Those in one group were prone to being punctual, while those in the other group were perpetually late. All of the volunteers were asked to judge how long it took for one minute to elapse. The punctual types came in pretty much right on time, while the latecomers were closer to the 80-second mark. Finally, conscientious people understand that the best-laid plans of mice and men often go awry, and make allowances for possible problems, such as buses running late or them having to struggle through a crowded street.

– Be realistic about how long it will take you to get somewhere. Allow for unexpected delays, wear a watch, and plan to arrive early. As some of the mission controllers used to say: 'If you are not 10 minutes early for a meeting, you are late.'

Habit 4: Adopt a frog-based diet

Conscientious people are especially likely to follow Mark Twain's advice and adopt a frog-based diet. Twain once remarked: 'If it's your job to eat a frog, it's best to do it first thing in the morning. And if it's your job to eat two frogs, it's best to eat the biggest one first.' Twain was describing a great way of promoting productivity. If there's something that you don't want to do, do it first thing in the morning, because you'll have more energy then, and it will provide a sense of accomplishment and momentum for the rest of the day.

– Start your day by confronting the hard tasks first.

Habit 5: Don't overcommit

One of the most important reasons why conscientious people keep their promises is that they don't overcommit themselves. In 2008, Emily Pronin of Princeton University presented volunteers with a horrible tasting concoction of soy sauce and ketchup.[120] Some of the volunteers were asked to decide how much of the unpleasant drink they were prepared to consume right there and then, while the others were asked how much they would be prepared to down in two weeks' time. The volunteers in the 'right there and then' group were willing only to drink two tablespoons, but those that were estimating how much they would down in the future said they would be prepared to drink half a cup. The same happens in everyday life. We all tend to overestimate how much time and energy we have in the future, and so end up taking on more than we can handle.

– When it comes to committing yourself to something in the future, think like a conscientious person by asking yourself; 'Would I want to do it tomorrow?' If the answer is 'no', find a way of politely declining the request.

Habit 6: Press the pause button

An inability to overcome the need for instant gratification can lead to poor habits, financial troubles, health issues, a lack of productivity, and all-round laziness. Conscientious people are good at avoiding temptation. When it comes to money, they tend not to buy stuff on a whim, exceed their credit limit or miss a bill payment. Similarly, when it comes to healthy eating, they don't tend to succumb to temptation and so are able to avoid chomping away on sweets and chocolates. Because of this high level of self-control, they find it much easier to prevent problems before they begin. For instance, when it comes to their finances, they don't end up buying something they don't want or need, they avoid paying late fees or getting a poor credit rating. Similarly, avoiding sugary snacks means that they are less likely to become overweight and suffer from several health issues.

– Overcoming the need for instant gratification often involves taking time out. When you are tempted to act on an urge, pause. Try to put some space between the moment of temptation and the moment

of action. For instance, if you're tempted to buy something on a whim, ask yourself if you really need to make the purchase. Even if the answer is 'yes', walk away, have a cup of coffee and think on it. Similarly, if you suddenly feel like reaching for an unhealthy snack, take a moment to reflect. Ask yourself whether this is the best way forward and, in doing so, give yourself an opportunity to make a much better choice.

Habit 7: Be fair

A few years ago, researchers at the University of South Florida asked a group of science students to complete a personality test and also report how much time they spent working in the laboratory.[121] Before completing the questionnaire, the students were told that they would receive a reward (course credits) for spending time in the laboratory, and the longer they said that they spent in the lab, the bigger the reward. The experimenters then secretly recorded how much time the students *actually* spent in the laboratory and discovered that the conscientious students had been far more honest than the other volunteers. This sense of honesty extends to everyday life. Conscientious people are less likely to cheat on their time sheets, steal office supplies, lie, break the rules in games and sports, litter, borrow something and never return it or use someone else's stuff without asking. They have a strong sense of fairness, are especially likely to apologize when they make a mistake,[122] and treat others with respect.

– Conscientious people are honest and as a result gain the trust and respect of those around them. Be honest and dependable. Don't exploit others, or take more than your fair share.

The Inflatable School

There's an old joke along the lines of: 'What did the inflatable teacher say to the inflatable student who ran around his inflatable school causing mayhem with a pin? You have let me down, you have let the school down and, most important of all, you have let yourself down.'

More seriously, not letting people down can help boost conscientiousness. Chris Kraft's remarkable style of leadership helped imbue Mission Control with the all-important 'It won't fail because of me' attitude. Kraft gave his staff a job to do and left them to it. Everyone respected him and worked hard, eager to show that they were worthy of his trust.

– Find your own Chris Kraft. Who do you admire in life? Maybe it's a parent, a boss, a colleague, or a friend. Now imagine doing whatever it is you are trying to do in order to impress them. Think about how you don't want to let them down and make sure that you carry out your duties to the best of your abilities.

– If you want to inspire others, try adopting Kraft's approach and give them a reputation to live up to and a desire not to let you down. As the great German writer and philosopher Johann Wolfgang von Goethe remarked: 'If you treat an individual as he is, he will remain how he is. But if you treat him as if he were what he ought to be and could be, he will become what he ought to be and could be.'

SUMMARY

When it comes to developing the attitude that gives you altitude, remember the Apollo mantra: 'It won't fail because of me.'

– Develop an internal locus of control by focusing on the power that you have to shape your life.

– Beat procrastination by thinking about the future, building a house one brick at a time and creating smart deadlines.

– Adopt the 7 habits of highly conscientious people by, for instance, adopting a frog-based diet, always arriving ten minutes early and asking yourself the magic question: 'Would I want to do it tomorrow?'

– Think about someone that you admire and imagine that they have trusted you to carry out some work. Don't let them down.

Moonshot Memo
Subject: Dedication

Here's a quick story to illustrate just how dedicated some of the mission controllers were.

Flight Controller John Llewellyn was a larger-than-life character. On one occasion he was driving to Mission Control when his car skidded off the road and went through a barbed wire fence. Llewellyn couldn't remove the car from the muddy field; instead he walked for miles in the dark and rain to get into work.

On another occasion, he was late for a shift and, unable to find a parking space, drove his car up the steps of a building and parked right by the front door. As a result, Llewellyn had his car pass taken away.

Undeterred, a few days later he arrived for work on his horse and tied it up in the Mission Control parking lot.

When you decide to do something, be dedicated and stay in the saddle.

6

'IF YOU'RE GOING TO GO TO THE MOON, SOONER OR LATER YOU'VE GOT TO GO TO THE MOON'

Where we venture into the unknown with the first manned mission to the Moon, and learn how you can find the courage to stop talking and start doing.

In late 1968, the Soviets announced another significant success in the race for space. Their mysteriously named Zond 5 spacecraft had headed out into space, circled the Moon and safely returned to Earth. On board were a curious collection of creatures, including two tortoises, a colony of wine flies and a couple of mealworms. The Soviets had also placed a tape player on board and used it to broadcast a recording of a cosmonaut talking. Exactly as planned, the Americans picked up the recording and, for a few moments at least, thought that their arch-enemies had managed to send a human to the Moon.[123]

When the Zond 5 returned to the Earth, the Soviet scientists discovered that the animals had survived the trip and that the only evidence of any ill effects came from the tortoises, who had lost around 10% of their weight. Once again, America suddenly appeared to be behind the curve. Worse still, the CIA had intercepted a series of secret messages suggesting that the Soviets were now gearing up to send humans to the Moon before the end of 1968.

The bad news kept coming. America's next space mission, Apollo 8, had originally been designed to orbit the Earth and test the Lunar Module. However, the Module was proving problematic to perfect, and engineers thought that it wouldn't be up and running until at least February 1969. Panicked by the success of the Zond 5, and worried about the delays in the Lunar Module, America's senior space officials came up with a bold, brave and remarkably agile change of plan. Instead of having Apollo 8 just zoom around the Earth, the officials wondered whether the astronauts could become the first humans to orbit the Moon.

FACING FEAR AND UNCERTAINTY

Apollo engineers and scientists were asked whether they could prepare for this new and ambitious mission in just a few months. The risks were obvious. If the spacecraft's trajectory was even slightly askew, the astronauts would either head off into deep space or crash into the Moon. In addition, the mission would take the astronauts behind the Moon, and so out of contact with Mission Control. During this time, they would have to carry out a tricky manoeuvre that was designed to place them into orbit around the Moon. Even the smallest of mistakes could prove fatal. But as

Flight Director Glynn Lunney described in his response to hearing the proposed change of plan, there was a sense of hey, why not:

> When I found out what they were thinking of doing, my first reaction was 'Wait a minute, we can't do that yet'. And then, when I thought about it for a minute or two, I said to myself 'Actually, that's brilliant, why didn't I think of that?' That was the reaction of lots of people. We had been preparing for this mission for years, and had intended to do it later in the schedule. Were we going to get smarter in a few months? No. Was the equipment going to get any better? Probably not. We knew what the risks were, and that, sooner or later, we were going to have to take those risks. What were we waiting for?[124]

At one meeting, Lunney summed up his thoughts in a single, powerful phrase: 'If you're going to go to the Moon, sooner or later you've got to go to the Moon.'[125] Mission Control agreed to the radical change of plan and astronauts Frank Borman, James Lovell and William Anders were chosen for the mission. The team was far from certain that the mission would be a success. Later on, Borman's wife Susan would ask Flight Director Chris Kraft for a realistic estimate of her husband coming home alive. After a short pause, Kraft replied: 'Fifty–fifty.'[126]

The game was afoot and everyone started to prepare for Apollo 8's historic flight. Each Apollo mission was an expensive affair. Sending a Saturn V rocket into space cost an estimated $185 million (around $1.16 billion in today's money), with the vehicle alone costing about $110 million. Assembly of Apollo 8's Saturn V was completed in late September 1968 and the giant rocket was slowly rolled out onto its launch pad about two weeks later. After extensive testing, the stage was set for lift-off on 21 December.

On the day before their departure, the Apollo 8 astronauts were visited by the world-renowned aviator Charles Lindbergh. In 1927, Lindbergh had achieved international fame by making the first non-stop transatlantic flight, covering the 5,800 kilometres between New York and Paris in 33 hours and 30 minutes. Now, just 41 years later, the Apollo 8 astronauts aimed to journey over 380,000 kilometres across space, orbit the Moon and return safely to the Earth. Just before leaving, Lindbergh calculated that the first one second of the Saturn V lift-off would use ten times more fuel than he had used on his entire transatlantic trip.

On launch day, hundreds of thousands of people gathered at the Cape, waiting for another gigantic Saturn V rocket to lift off from the Kennedy Space Center. A few miles away, Borman and his crew entered the White Room, went through their final preparations and climbed into their Command Module. The ever-attentive Günter Wendt had hung a small Christmas decoration on each of the module's seats.[127] Wendt carefully tightened the astronauts' seat belts and then closed the Command Module's hatch. Finally, Apollo 8 was ready to blast into space. Each stage of the Saturn V rocket performed perfectly and soon the crew was in orbit around the Earth. Three hours later the astronauts fired up their engine, accelerated to around 38,000 kilometres per hour and became the first humans to begin to journey to the Moon.

There was a great deal to do. A major part of Apollo 8's navigation system relied on star maps, a telescope and a sextant. This was the same type of equipment that Columbus had used to make his way to the Americas and the crew regularly had to use this age-old technology to ensure that they were heading in the right direction. In addition, the astronauts also had to rotate their craft on an hourly basis to prevent the Sun from overheating the outside panels – a procedure known as a Passive Thermal Control or, as most of the team referred to it, a 'barbecue roll'.

Unfortunately, the crew also had to deal with their urine, faeces and vomit. During the journey, Borman suffered from severe 'space-adaptation syndrome', resulting in the craft becoming full of small globules of vomit and diarrhoea. In an attempt to minimize the disruption, the astronauts were forced to scramble around the cabin trying to capture these blobs in paper towels.

The other negative was that the food on Apollo 8 wasn't great. The astronauts' meals had been frozen and then placed into a vacuum chamber to remove all of the moisture. When the astronauts became peckish, they used a water gun to inject either hot or cold water into the package and then squeezed the resulting mush into their mouths.

All went surprisingly well: after three days of vacuum-packed meals, barbecue rolls and floating faeces, Apollo 8 was fast approaching the Moon.

'THE CUSTARD IS IN THE OVEN AT 350': HOW TO KEEP CALM AND CARRY ON

To begin orbiting the Moon, the crew had to carry out an especially dangerous and tricky manoeuvre. The procedure involved rotating their spacecraft around and then firing their engine system. Producing a force in the opposite direction to travel, this was designed to slow their spacecraft down and ensure that it was captured by the Moon's gravity. If the engine system failed to activate, then the crew would be slung around the Moon and sent straight back to the Earth. If the engine burn went on for too long, the crew ran the risk of rapidly descending to the lunar surface and inadvertently becoming the first humans on the Moon. Worse still, this difficult procedure had to happen when the astronauts were behind the Moon and therefore out of contact with Mission Control.

About three minutes before this loss of contact, Mission Control sent the astronauts a cryptic, custard-based communication. This strange message had its origins with Apollo 8 Commander Frank Borman and his wife Susan. Frank had served in the military and worked as a test pilot. When Susan was aware that her husband was facing danger, she used the phrase 'the custard is in the oven at 350', as a light-hearted way of saying that she was thinking about him and looking after their family. Some of the mission controllers knew about the phrase and around three minutes before Apollo 8 disappeared behind the Moon, Mission Control messaged Borman:

'The custard is in the oven at 350. Over.'

It was a lovely thought and a touching gesture. Unfortunately, the unexpected nature of the communication, combined with high levels of static noise, resulted in Borman struggling to understand the line. Moments later, Borman replied, 'No comprendo'. And that was that.

A few seconds away from loss of contact, Mission Control wished the crew a safe journey. Lovell optimistically responded: 'We'll see you on the other side'; moments later, Apollo 8 vanished behind the Moon.

The astronauts now had to carry out the manoeuvre that would place them in orbit around the Moon. There was little room for error. In an ideal world, they would turn their spacecraft around and then, at exactly the right moment, ignite their engine for precisely 4 minutes and 2 seconds. If everything went to plan, Mission Control would resume contact with Apollo 8 around 32 minutes after it had vanished from their screens.[128]

The mission controllers waited in near silence. Everyone knew that if the craft didn't emerge, they would have no way of knowing what had gone wrong. Under those circumstances, it would be foolhardy to send another crew on a similar mission, possibly marking the end of the entire Apollo programme. Realizing that everyone could do nothing but wait, Flight Director Glynn Lunney informed the controllers that now was a good time for a comfort break. Cigarettes were lit, gum was chewed and coffee was sipped.

REAPING THE BENEFITS OF BRAVERY

The engine burn proceeded exactly as planned, with the astronauts later describing it as the longest 4 minutes of their lives. Apollo 8 entered lunar orbit, Borman, Lovell and Anders becoming the first humans to set eyes on the far side of the Moon. Then, right on cue, Apollo 8 reappeared on the Mission Control screens and resumed contact with the Earth. The controllers cheered, clapped and congratulated one another. The good news was broadcast to a waiting world and millions of people joined in the celebrations. In Britain, the country's leading astronomer, Sir Bernard Lovell, described the moment as 'one of the most historic developments in the history of the human race'.[129]

Orbiting just over 100 kilometres above the Moon, the three astronauts could see the lunar surface close-up. Peering out through the tiny windows in their spacecraft, all three were awestruck by the seemingly endless sea of greyish moon dust, with Borman later describing it as 'a vast, lonely, forbidding expanse of nothing'. For the next twenty hours, the three astronauts repeatedly circled the Moon, taking photographs, assessing possible landing sites for forthcoming missions and naming craters. In a touching gesture, Jim Lovell dedicated 'Mount Marilyn' to his wife.

At one point, the Earth emerged from behind the lunar horizon, and the crew became the first humans to witness an 'Earthrise'. Excited by what they were seeing (Anders: 'Oh my God! Look at that picture over there! There's the Earth coming up. Wow, is that pretty'), they picked up a camera and took several photographs of the remarkable scene. In 2016, *Time* magazine chose the best of the photographs as one of the most influential and poetic images of the century.

On Christmas Eve, Apollo 8 circled the Moon for the ninth time and the crew began a live television transmission to Earth. Watched by millions of viewers, each of the astronauts read a section from the Book of Genesis ('In the beginning, God created Heaven and Earth . . .'), and Borman ended the broadcast by wishing the people of Earth a Merry Christmas. The idea for the reading had originated with the wife of a NASA public affairs officer and the text had been typed on fireproof paper and inserted into the back of the crew's mission notes.[130] The transmission proved to be a poignant and dramatic moment and would later result in the astronauts being given an Emmy Award. Many of the mission controllers were people of faith and the Biblical reading was especially significant for them. Sitting at the front of Mission Control, Jerry Bostick quietly said a prayer and thanked God for letting him be part of this historic enterprise.[131]

While many of their fellow Earthlings took a well-earned Christmas day nap, the Apollo 8 crew prepared for another highly dangerous task. To head back to their home planet, the astronauts had to fire up their engine again. If the engine system failed to activate, the crew would be trapped in lunar orbit and eventually die circling the Moon. Similarly, if the burn went on too long, or wasn't carried out at precisely the right moment, Apollo 8 could shoot off into deep space. Once again, this tricky procedure had to be carried out while the astronauts were on the far side of the Moon and therefore out of contact with Mission Control.

Charles Deiterich was in Mission Control that day. Like so many of his colleagues, Deiterich came from a modest background. Born in a small town in Pennsylvania, his father was a machinist and his mother was a teacher at a rural school. As a young man he obtained a university scholarship to study physics, launched several homemade rockets and was eventually offered a position at NASA's Manned Spacecraft Center. During the Apollo 8 mission, Deiterich helped ensure that astronauts made it safely back to the Earth. He remembers the period well:

It was an enormous amount of work. The first floor of the building that housed Mission Control contained several large mainframe computers. When we wanted the spacecraft to take a certain trajectory, we would send a request down, and the computers would come up with the data that we needed to make it happen. We had a lot of that information prior to the mission, and so we could tell the astronauts how long the engine burn should be, when it should happen, and which stars to look out for in their telescope so that they knew they were pointing in the right direction. Before Apollo 8, I can remember meeting Frank Borman and using his shaving foam dispenser to illustrate how the spacecraft needed to be positioned during re-entry![132]

On Christmas Day 1968, Apollo 8 vanished behind the Moon and attempted to carry out the engine burn that they hoped would take them back towards the Earth. Back in Mission Control, Deiterich and his colleagues waited to discover whether their calculations were correct. Around forty minutes after losing contact with Apollo 8, Mission Control attempted to regain contact with the astronauts. Their initial attempts were met with a stony silence. Then suddenly the consoles lit back up as the spacecraft appeared on their screens. The burn had proceeded like clockwork and Apollo 8 was heading home. To Lovell, the homeward bound trajectory was a wonderful Christmas present and he shared his sense of joy with Mission Control: 'Please be informed, there is a Santa Claus.'

A few hours later the crew opened their food locker and were delighted to find a small parcel tied up with red and green ribbons. Inside was a pack of thermostabilized turkey, freeze dehydrated cranberry-apple sauce and three miniature bottles of brandy. Borman ordered his crew to abstain from the brandy and the bottles remained unopened for many years after the flight.

Moonshot Memo
Subject: *'The most remarkable thing about the whole team was how unremarkable we all were!'*

Listen to recordings of some of the most historic moments in the Apollo programme and you are likely to hear Doug Ward. Known as the 'voice of Apollo', Ward broadcast live from Mission Control and provided the public with a cool-headed account of the often dramatic events that were unfolding in front of him. All these years on, Ward still has a vivid memory of what it was like to work in Mission Control:

> I would invariably come into the control center having walked across the Manned Spacecraft Center campus in blinding Texas sunlight and insufferable Houston humidity. Inside, the heating and air conditioning systems maintained the cool temperature and low humidity necessary to keep the big IBM computers happy. Light levels were often very low to make console and front screen displays easily visible. Others have commented on smells of pizza, coffee and cigarettes in Mission Control, but I was mainly struck by the sense of quiet calm and cool darkness.[133]

Ward considers Apollo 8 to be perhaps the most significant of the Apollo missions. He provided the live commentary for when the astronauts carried out the manoeuvre that everyone hoped would bring them home:

The engine burn took place on the far side of the Moon, out of radio contact with the Earth. If the burn was successful, they would re-establish contact a few seconds earlier than if it hadn't gone according to plan. I was sitting next to Chris Kraft and he was usually very cool headed, but I could tell he was very tense as we waited to discover if the astronauts were safely headed home. Two clocks counted down to re-acquisition of a signal from the spacecraft. One of the clocks indicated if the burn had been successful and the other indicated if it had been unsuccessful.[134]

Ward (25 December 1968): We now show less than 30 seconds until re-acquistion. We will stand by for the first words from the Apollo 8 crew as they come over the lunar horizon, and into acquisition.

As the first clock reached zero, Mission Control radioed the crew again and again ... and received no response ... and then we picked up data from the spacecraft.

Ward (25 December 1968): There is a little bit of a cheer going up among the flight controllers here. We should be hearing from the crew shortly.

And that's when Jim Lovell came on and said: 'Pleased be informed, there is a Santa Claus.' There was a tremendous sense of relief from everyone, especially Kraft.

Like so many of those in Mission Control, Ward came from a modest background and was passionate about his work:

I was the son of a hardware store manager and a stay-at-home mother. I loved broadcasting and was working as a disc jockey on my local radio station when I was still in high school! Looking back, I guess the most remarkable thing about the whole team was how unremarkable we all were![135]

As they approached the Earth, the Apollo 8 crew jettisoned the Service Module that had provided them with food, water and power. There was no going back now. When the astronauts re-entered the Earth's atmosphere, their spacecraft travelled at a breathtaking 40,000 kilometres an hour and was transformed into a fireball burning at around 5,000 degrees Fahrenheit. Fortunately, their heat shield worked perfectly and Apollo 8 eventually splashed down on 27 December 1968.

It had been a difficult year for America. The Viet Cong and North Vietnamese soldiers had caught American forces by surprise during the Tet Offensive. Both Martin Luther King Jr. and Robert F. Kennedy had been assassinated, and much of the country was filled with political unrest. Against this backdrop of misery, anguish and introspection, many people saw the success of the Apollo 8 mission as a beacon of hope. *Time* magazine announced that Borman, Lovell and Anders were to be its 'Men of the Year' and the three astronauts received thousands of letters of support. Perhaps the most symbolic telegram was sent to Borman. Anonymous, it simply read: 'Thank you Apollo 8. You saved 1968.'

Apollo 8 was the first time that humans had travelled hundreds of thousands of kilometres across space, and the first time that anyone had seen the far side of the Moon in person. It required an enormous amount of planning and a great deal of courage. But then, as Glynn Lunney memorably put it: 'If you're going to go to the Moon, sooner or later you've got to go to the Moon.'

*

Finding the Courage to Stop Talking and Start Acting

'Whatever you can do. Or dream you can do. Begin it. Boldness has genius, power and magic in it. Begin it now.'
— Goethe

Launching Apollo 8 was a courageous decision. The mission was clearly highly risky, and the Apollo team had a relatively small amount of time to prepare. However, Mission Control realized that the time had come to stop talking and start doing. The risks paid off and the mission was a phenomenal success.

You probably won't have to decide whether to launch a mission to the Moon, but may well be faced with many other equally scary decisions. Maybe you are thinking about walking away from an unhappy relationship, but are fearful of being alone. Or you are considering leaving your tedious job and starting to work for yourself, but are afraid about the potential financial uncertainty. Or perhaps you want to write a novel, but are concerned about the possibility of people being unimpressed with your writing. Or maybe you are thinking about starting a new project at work, but are worried that it might end in failure.

For over fifty years, psychologists have studied how people

respond to these types of daunting decisions, and discovered that their responses fall into one of two general categories: Flight or Fight.[136]

The differences between the two groups can be illustrated with a simple scenario. Let's imagine that you are considering leaving your safe but dull job and setting up as a freelancer. There are clearly considerable uncertainties and financial risks involved in being self-employed, and so, quite understandably, you find the idea somewhat scary.

If you have a 'flight' mentality, then you will have a tendency to run away from your fears. This approach is often associated with being risk-averse and valuing the short-term comfort of the status quo over the uncertainty associated with long-term change. In our imaginary scenario, this would involve you focusing on the certainty of losing your comfortable source of income and the potential risks associated with self-employment. It doesn't stop there. Deep down, you would know that you are scared to take the plunge, but save face by making up various excuses. You might, for instance, tell yourself and others that you are sticking with your safe job until the economic climate has improved.

In general, this way of coping with fear is associated with failure. Running away from a threat increases the likelihood of becoming trapped in a suboptimal situation, and makes you more likely to feel unhappy, afraid and unfulfilled.

In contrast, you might be far more fight than flight and have the courage to face your fear in the hope of bringing about a better future. This approach is associated with being action-based, risk-taking and focusing on overcoming potential problems. In our imaginary scenario, you might be scared about the possibility of leaving your safe job, but are prepared to assess the risks associated with both making change and sticking with the status quo. Without being reckless, you would be more likely to tolerate the uncertainties associated with being your own boss for the potential prize of taking control of your life. If you do make the decision to leave, you will jump into action, rather than finding excuses to procrastinate and delay.

This action-based approach results in two major advantages. First, you learn by doing. While your flight-minded friends are busy talking the talk, you are rolling up your sleeves and getting on with it and so are far more likely to develop the skills needed to make your plans a reality. Second, by putting yourself out there, you increase the likelihood of meeting other like-minded people and coming across unexpected opportunities.

Perhaps not surprisingly, the fight mentality is associated with success because it makes you less afraid, encourages action and helps you to grow and to develop.

The good news is that there are several simple techniques to help you become more fight than flight. These techniques help you to properly assess the risks involved in staying or shifting, to encourage you to stop talking and start acting, to stop you being overly reckless, and to motivate you to overcome the unpleasant bodily sensations associated with fear.

RISKY OR RECKLESS?

How nervous do you feel when you step onto an aeroplane? And how about when you are driving home? Perhaps not surprisingly, most people report feeling far more scared when they are flying than driving. In reality, you are about a hundred times more likely to die during the drive than during the flight. Not only that, but when most people do arrive home, they are faced with several death traps, including slippery rugs at the top of their stairs, dodgy wiring carrying hundreds of volts and sharp knives placed precariously on kitchen counters. No wonder then, that statistics show that your home is far from sweet and that thousands of people die in household accidents each year.

This illogical approach towards risk becomes even worse when you add fear into the mix. When people become afraid, they tend to focus on self-preservation, develop an aversion to risk, and choose the safest short-term option. Given that change usually carries some sense of uncertainty, this often results in them sticking with the status quo. Scared of losing their secure salary, they remain in a dull and unrewarding job. Afraid of being alone, they stay in an unhappy relationship. Worried about failure, they decide not to embark on a new project.

The following exercise is inspired by a technique described by Tim Ferriss in his book *Tools Of Titans*. It is designed to bypass the damaging effects of fear and help assess the upside and downside of a decision in a more cool-headed and rational way.

1) Think about a decision that makes you feel frightened or anxious. It probably won't involve you deciding whether to launch a mission to the Moon, but might, for instance, involve a change in your career, asking someone out on a date, finishing a relationship or starting a new project.

   The decision that I have in mind is: _____
   _____

2) Imagine that you have a rose-tinted crystal ball and are able to see your life a year from now. Stare deep into the ball. Imagine that you took the plunge and made the change, and everything turned out surprisingly well. Spend a few moments describing the best-case scenario that comes to mind.

The best-case scenario is:_____

_____

3) But hold on. The crystal ball is starting to cloud over, and a very different future is emerging. You are seeing your life in a year's time. You took the plunge and made the change, but this time it was a complete disaster. Spend a few moments noting down the worst-case scenario that you can imagine.

The worst-case scenario is:_____

_____

4) Next, please rate how likely you think the worst-case scenario actually is, on a scale from '1' (very unlikely) to '10' (very likely). Don't think about it too much. Try to give an honest assessment.

I believe that the likelihood of the worst-case scenario is:

_____

5) Now, ask yourself how you could cope if the worst-case scenario became a reality. Would it really be that terrible? What would you do to repair the damage? How have you coped with sim-

ilar problems in the past? Who could you ask for help? Have other people been in the same situation and pulled through? Spend a few moments jotting down your thoughts.

I would try to cope with the worst-case scenario by: ____

_____

6) Next it's time to think about whether there is anything that you can do to make this worst-case scenario less likely to happen or to minimize the impact. What precautions could you take? Do you have the skills to help prevent the worst-case scenario or could you develop them? Again, spend a few moments jotting down your thoughts.

I would try to prevent the worst-case scenario by: _____

_____

7) When making an important decision, people often focus on the risks associated with doing something and ignore the costs associated with inaction. Imagine that the status quo prevails and that you don't do anything at all. Again, get out your crystal ball and look into the future. If you do nothing, what does your life look like in a year's time?

The risks associated with the status quo are: _____

_____

This exercise is designed to help you figure out when to listen to your fears and when to ignore them. Take a careful look at your answers. What are the benefits of everything going really well? Does that future excite you? What's the worst that could happen, and how would you cope? How likely is that disastrous scenario and what can you do to help prevent it happening? How does that make you feel? And what are the costs associated with not doing anything at all? Remember, as President John F. Kennedy once remarked: 'There are risks and costs to action. But they are far less than the long-range risks of comfortable inaction.'

You might conclude that your fear was well founded and that you are better off sticking with the status quo. That's fine. Alternatively, you might decide that the best-case scenario is a prize worth fighting for, that you can cope with the worst-case scenario or that the costs associated with inaction are unacceptably high. Under these circumstances you might decide to find the courage to face your fear and choose to change. And that's fine, too.

Most self-development gurus will encourage you to feel the fear and do it anyway. In fact, it is much better to develop a more rational and cool-headed insight into the best way forward, to take risks but not be reckless.

IF NOT NOW, THEN WHEN

Moving from a flight-to-fight mentality can be surprisingly tricky. Even when you have made the decision to confront your fear and step into the unknown, you may still face several psychological hurdles. As you stand on the cliff edge of change, your brain may suddenly tell you not to jump. Rather than acknowledge that you are scared, you might be tempted to start to make excuses to yourself and others. This exercise is about tackling the four most common types of excuses that people use to justify inactivity.

If you find yourself using these excuses when you speak to yourself and others, ask the follow-up questions to help discover whether your delay reflects a genuine worry or is simply you giving in to your fears and concerns.

I want to make the change but . . .

I simply don't have the time.

How can you find the extra time that you need? What would happen if you changed your priorities and put your new plan or project towards the top of your list?

. . . I don't have the money, information or skills yet.

Do you really need these resources to make a start? If you do, is it possible to sell something to raise the cash? Or to find out the necessary information? Plans do not need to be perfect to make a start. If you have 70% of what you need, go for it.

. . . I am waiting for exactly the right time to launch into action.

That might be fine, but be careful of paralysis by analysis. Are you simply avoiding the moment? Try setting yourself a hard deadline for action.

. . . people like me don't tend to succeed, so there's no point in trying.

Are you telling yourself that you are from the wrong background or didn't have the right upbringing? These types of factors can't be changed and so can feel especially comforting when it comes to justifying a lack of action. Can you find people who are just like you, and succeeded?

Finally, be wary if you find yourself starting a project, but never getting around to completing it. Again, this might be fine, but could be a sign of a classic 'if I never finish then I will never fail' attitude. Can you show people what you have achieved so far (often fear of failure involves people saying that their work must be perfect before they will show it to anyone)? When do you think you will finish?

It's natural to feel like you should avoid whatever it is that makes you scared. However, it's important that you work through this tricky time, because the longer you give in to fear, the stronger the fear will grow. Skip the excuses, stop talking and start doing.

WE DON'T HAVE TO GO TO THE MOON TODAY

The future is uncertain and once you start a new project it may not go according to plan. Although persevering in the face of adversity is important, it's equally vital that you don't become so invested in a scheme that you keep on going when it would be more sensible to take a break and try another time. The Apollo programme provides a perfect example of the best way forward.

Gerry Griffin was the Flight Director for the Apollo 12 mission. On 14 November 1969, the Saturn V rocket carrying the three Apollo 12 astronauts lifted off from the Kennedy Space Center. Initially, all went well. But then, thirty seconds into the launch, all hell let loose. The astronauts saw a bright flash of light illuminate their capsule and, moments later, heard a blast of static noise in their headphones. Suddenly, major alarm signals started to sound and their instrument panel was awash with red and yellow warning lights.

Proving the old adage wrong, the Apollo 12 Saturn V rocket had just been hit by two strikes of lightning. Back in Mission Control, the data screens were full of jumbled numbers and Griffin had seconds to make a life-or-death decision. Either he could waste millions of dollars by aborting the flight or allow the mission to carry on and place the astronauts into a potentially fatal situation. Suddenly, 24-year-old Flight Controller John Aaron suggested that having the astronaut's throw a single switch ('SCE to Aux') would probably bring the data back online and buy Griffin some much-needed extra time. His hunch proved correct and the warning lights slowly started to vanish.

With the immediate crisis averted, Griffin and his colleagues tried to figure out just how badly the craft had been damaged by the lightning strikes.

'I will never forget that moment,' recalled Griffin, 'We were trying to decide whether to go on to the Moon, and my boss, Chris Kraft, came over to me and gently said "Young man, we don't have to go to the Moon today."'[137]

'Kraft was saying two things', added Griffin, 'First, none of the more senior people there were going to step in. I had their trust and this was up to me. Second, it was a gentle reminder that I shouldn't get "go-fever" and feel the need to push on with the mission if I thought it was too risky.'

A vast amount of time, energy, and money had been invested in the mission. Nevertheless, Kraft told Griffin that he shouldn't feel pressured into making a reckless decision. It was also another striking example of the way in which the Apollo leadership placed their trust in people and imbued them with a sense of responsibility.

Griffin decided that it was safe to continue the mission and was eventually proved right. Apollo 12 was a huge success, with the crew returning safely to the Earth ten days later.

If a new project isn't going well, be careful not to make a bad situation worse by becoming reckless and continuing to invest time, energy and finances. Perhaps it's best to walk away, and try again another day. As Kraft told Griffin, maybe we don't need to go to the Moon today.

Moonshot Memo
Subject: The importance of 'today'

Quick thought: when Kraft said 'We don't have to go to the Moon today', the word 'today' was important. He wasn't saying that they would never go to the Moon again, but merely that Griffin shouldn't feel any pressure to push ahead there and then. If you do decide not to head to your Moon, leave the door to future opportunities open by adding the word 'today'.

Walking towards the Cannons

When you hear the sound of the cannons, walk toward them

A few months after President Kennedy declared that America would put a man on the Moon, MGM released a romantic comedy film entitled *The Courtship of Eddie's Father*. Based on a popular novel by Mark Toby, a key part of the plot revolved around the Henrietta Rockefeller Poise and Confidence School – a fictional organization devoted to betterment and personal growth. Attendees were offered several rules to increase their likelihood of success in life, with the first and most important rule involving doing one thing every day that scared them.

174

This wasn't the first time that Americans had been encouraged to regularly embrace their fears. In his 1841 essay, 'Heroism', Ralph Waldo Emerson urged his readers to 'Always do what you are afraid to do'. And in 1960, Eleanor Roosevelt's book *You Learn by Living* contained the same advice, with the First Lady noting:

'Fear has always seemed to me to be the worst stumbling block which anyone has to face . . . You gain strength, courage, and confidence by every experience in which you really stop to look fear in the face . . . You must make yourself succeed every time. You must do the thing you think you cannot do.'

The notion has stood the test of time. In 1997, *Chicago Tribune* columnist Mary Schmich offered her readers a series of life lessons, including, 'Do one thing every day that scares you.' Schmich's list quickly went viral and eventually formed the basis for the hit song, 'Everybody's Free (To Wear Sunscreen)'.

The advice has stood the test of time because it works. In the same way that psychologists help people overcome phobias by exposing them to the very thing that they fear, so doing something that scares you makes you more courageous and confident.

Obviously, it shouldn't be anything that is truly reckless. Jumping off the top of a high building is terrifying for a good reason and should be avoided at all costs. However, carrying out a much safer, but nevertheless equally scary, task is good for the soul. Here is a list of ideas that you might want to consider. See how you feel when you read each one. If an idea suddenly gives you butterflies in your stomach, see if you can find the courage to face your fears.

– Pose nude for a drawing class.

– Go rock climbing, caving, skydiving, or zip-lining.

– Visit a restaurant that serves insects, and eat a cricket, grasshopper or scorpion.

– Go to a party or networking event by yourself and introduce yourself to five people that you don't know.

– Remove all of your clothing and wander around your house completely naked.

– Think about something that you are not very good at or that frightens you (ballet, learning to swim, mastering a new language or the trapeze), and sign up for a class on the subject.

– Change your hairstyle. If you have long hair, cut it short. If you have short hair, let it grow long. Maybe dye your hair a very different colour.

– Tell your parents, partner or closest friend how you really feel about them (assuming it's positive!).

– Take a ride in a helicopter or on a roller coaster.

– Tell the truth when it would be much easier to lie.

– Get a small tattoo or a piercing.

– If you are afraid of chatting to people that you don't know, go to the park and tell a stranger that you think their dog is adorable.

– Spend the weekend disconnected from your smartphone, tablet and computer. No web browser, email or social media.

– Give something significant away. Maybe some money to a deserving cause that you don't normally support or one of your favourite possessions to a friend or family member.

– If you are afraid of standing up in front of people, arrange to give a talk or even consider performing a short stand-up routine at the try-out night of your local comedy club.

– Go on a significant trip alone (again, be safe).

– Go on a ghost walk or spend the night in a house that is allegedly haunted.

Moonshot Memo
Subject: Exposing yourself

Astronauts are often amazingly calm in potentially scary situations, in part because they are trained to realize that panicking doesn't change or improve the situation.[138]

Although you probably won't have to face the trials and tribulations of blasting off into space, there might be something that freaks you out. People have phobias about many different things, including spiders, speaking in public, flying, chopsticks and clowns. Years of research have revealed that one of the best ways to overcome a phobia involves repeatedly facing your fear. This technique, known as 'exposure therapy', is systematic and slow.

Let's imagine that you are afraid of spiders. You might start off by relaxing and then looking at a photograph of a spider from across the room. Repeat the experience a few times and you will find that the distant encounter no longer makes you feel anxious. Once that happens you might repeat the process, but this time moving closer to the photograph. After a while, you will start to feel fine looking at the photograph and be able to move on to the next stage. Over time, you move on to increasingly close calls, including, for instance, looking at a real spider in a box, holding the box and finally handling the spider.

The same technique can eliminate almost any phobia. If you feel anxious in social situations, you might start off by saying hello to a supermarket cashier, then stop someone in the street and ask for directions, then make small talk with restaurant staff and finally go to a party where you only know a handful of people. Similarly, if you are afraid of clowns, you might start off by looking at a photograph of one, then being in the room with someone wearing overly large shoes and slowly working your way up to getting a large custard pie in the face.

In short, the more you get used to doing something, the less anxious you will feel when you do it again.

SUMMARY

Sometimes it's important to feel the fear and do it anyway. However, often this is easier said than done. To find out whether the time is right to stop talking and start acting, remember:

– Imagine the best- and worst-case scenarios. How likely is the worst-case scenario and what steps can be taken to avoid it? Is taking the plunge risky or reckless?

– It's easy to put off doing something that scares you by coming up with a series of excuses. You might tell yourself that the time isn't right or that you don't have everything you need to make a start. Ask yourself whether these are genuine reasons for inaction or excuses born of fear.

– Be careful of carrying on any enterprise because you have made a start or investment. Remember that you don't need to go to the Moon today. If a scheme is becoming too costly or reckless, take time to consider your options.

– Get used to doing things that scare you. When you have the opportunity, find the courage to walk towards the cannons.

7

THE MAN WHO SAID 'GO'

*Where we learn how quick thinking saved the historic
Moon landings with just seconds to spare, and find out
how you can prepare for almost every eventuality.*

Apollo 8 was a huge success and now, the Apollo team had just one year to meet Kennedy's deadline of having a person step onto the lunar surface before the decade was out. Very soon the Lunar Module was up and running, and both of the subsequent Apollo missions put this new spacecraft through its paces. Apollo 9 blasted off in March 1969, orbited the Earth, and rehearsed essential rendezvous and docking procedures. Everything went according to plan and, just two months later, Apollo 10 was up and away. Following in Apollo 8's footsteps, the Apollo 10 astronauts travelled across space and orbited the Moon. Once there, two of the astronauts climbed into the Lunar Module, descended to within a few kilometres of the Moon's surface, and then returned to the Command Module. The speed with which these two missions were planned and executed was breathtaking, and the mission controllers had little time to appreciate their achievements. As Apollo Flight Controller John Aaron once memorably phrased it, sometimes it felt as if they were gulping fine wine.[139]

The success of Apollo 10 meant that the dress rehearsals were over and the team were now ready to put a person on the Moon. The Apollo 11 crew consisted of Neil Armstrong, Buzz Aldrin and Michael Collins. The three men were very different to one another. Neil Armstrong designed model planes as a boy, obtained his pilot's licence before his driver's licence and flew almost eighty combat missions during the Korean War. Buzz Aldrin graduated third in his class from West Point, had carried out extensive research into the rendezvous of space vehicles and was the only Apollo astronaut to have a doctorate. His nickname was 'Dr Rendezvous'. In complete contrast to his two crewmates, Michael Collins liked to paint, was an avid gardener and wasn't especially interested in technology.

On the morning of 16 July 1969, Armstrong, Aldrin and Collins suited up and made their way over to the Apollo 11 Saturn V rocket. They took the lift up to the top of the rocket's support tower, entered the White Room and began to make their final preparations, as ever under the watchful eye of the eccentric and reliable Günter Wendt. As Armstrong clambered into the Command Module, Wendt handed him a playful parting gift. Wendt had fashioned a small crescent-shaped trinket from foil-coated styrofoam, and explained to Armstrong that it was the key to the Moon.[140] Armstrong thanked him, explained that he was pushed for space in the module and asked Wendt to hold on to it until he returned home. In return, Armstrong handed Wendt a mock space-taxi ticket that was 'good between any two planets'.

A few miles away, hundreds of thousands of people had flocked to the Cape to witness history in the making. NASA's official guest list alone ran to 20,000 names, including the veteran entertainer Johnny Carson, former President Lyndon B. Johnson and then-current Vice President Spiro Agnew. They were joined by 2,000 journalists representing over 50 countries. The highways were in complete gridlock and the local vendors quickly sold out of their T-shirts, caps, and badges.[141] Desperate for souvenirs, some people resorted to pulling up the grass and putting it into their bags.

At 9:32 a.m. Eastern Daylight Time, the Saturn V rocket blasted off from the Kennedy Space Center, rose majestically into the sky and slowly vanished into the clouds. As the shock waves hit the waiting crowds, everyone clapped, cheered, and hollered. Everything went according to plan and within a few minutes the three astronauts were orbiting the Earth at a speed over 28,000 kilometres per hour. After one and a half trips around their home planet, the crew fired up their engine and accelerated towards the Moon. The world's most ambitious space mission was on its way.

Over the years, the mission controllers had realized that the astronauts could become confused if too many people attempted to communicate with them during their journey. As a result, only one person in Mission Control was allowed to talk directly to the astronauts while they were in space. This position, known as 'CAPCOM', was always assigned to a fellow astronaut to ensure that the crew spoke to someone who understood the mission from their perspective. Throughout the Apollo 11 mission, several people acted as CAPCOM and provided Armstrong, Aldrin, and Collins with important information and news updates. Many of the communications were vital to the success of the mission. However, once in a while, there was room for more light-hearted banter. On 18 July, for instance, CAPCOM informed the Apollo 11 crew that Irishman John Coyle had just won the world porridge-eating championship by consuming twenty-three bowls of oatmeal in ten minutes. Up in space, Collins playfully suggested that Aldrin could enter next year's competition as he had just consumed his nineteenth bowl of oats.

Moonshot Memo
Subject: The F word

Lessons had been learned from Apollo 8 and none of the astronauts suffered any vomiting or diarrhoea during their three-day trip to the Moon. But that is not to say that the environment inside the Command Module was pleasant. The Apollo 11 fuel cells combined oxygen and hydrogen to produce both electric power and drinking water. Unfortunately, this resulted in the water containing lots of hydrogen bubbles which, in turn, caused the astronauts to produce a considerable amount of flatulence. Armstrong described the smell inside the Command Module as 'a cross between wet dog and marsh grass', and Aldrin once quipped that the flatulence could be used as an alternative propulsion system. Although Aldrin was joking, some space scientists were worried about the air inside the Command Module.

The air convection (warm air rising, cold air falling) that occurs on the Earth doesn't take place in the micro-gravity environment of space. Without artificial convection, the air in the Command Module would remain very still and this could cause several serious issues. For instance, when the astronauts were asleep, the carbon dioxide they exhaled could build up around them and lead to breathing difficulties. In addition, the lack of convection could result in equipment not being cooled by cold air and so quickly overheating. Perhaps most worrying of all, the astronauts' flatulence could become trapped in sections of the spacecraft and run the risk of exploding.

As a result, scientists and engineers had devised a highly efficient air conditioning system, created equipment that was cooled by circulating fluids, and carried out extensive research into the F-word (titles of the resulting academic papers include 'Space gastroenterology: A review of the physiology and pathology of the gastrointestinal tract as related to space flight conditions' and 'Intestinal hydrogen and methane of men fed space diet').[142] If any of this technology failed, the consequences could be fatal.

A few days after lift-off, Apollo 11 was fast approaching its destination. Just hours away from the Moon, CAPCOM McCandless informed the crew that West Germany had declared the following Monday (the day when Armstrong and Aldrin planned to land on the lunar surface) to be Apollo Day, that Bavarian children would be getting the day off school and that the Pope had arranged for a special colour television circuit to be installed in his summer residence to watch the great event (at the time, Italian television was broadcasting in black and white).

Exactly as planned, Apollo 11 passed behind the Moon and the astronauts successfully completed the tricky engine burn that placed them into lunar orbit. The stage was set for Armstrong and Aldrin to begin their historic journey down to the surface of the Moon.

THE ULTIMATE PEP TALK

On 20 July, Flight Director Gene Kranz arrived for work and walked into Mission Control.[143] He had been chosen to lead the Lunar landing phase of the mission, and knew that he was hours away from either fulfilling Kennedy's vision, being forced to abort the mission, or losing two brave men. CAPCOM Charlie Duke was calmly conveying the news and sports headlines to the astronauts.

Twenty-six-year-old Steve Bales was sitting behind one of the consoles in Mission Control that day. Like so many of his colleagues, Bales came from a modest background. The son of a school janitor and a beautician, Bales was raised in a small rural farming community in Iowa. When he was a teenager, he saw Wernher Von Braun on television talking about the joy of space, and became fascinated with the idea of sending astronauts to the Moon. After obtaining a degree in Aeronautical Engineering from Iowa State University, Bales was hired by NASA and originally gave tours around the Johnson Space Centre. He became friendly with several of the mission controllers and was eventually asked to work as a guidance officer in Mission Control. He had no idea that he was hours away from playing a hugely significant role in Armstrong's historic descent.

NASA had reverted to their original policy of allowing the astronauts to name their spacecraft, and the Apollo 11 Lunar Module had been christened 'Eagle' and the Command Module 'Columbia'. At around 9.30 a.m. Eastern Daylight Time, Aldrin and Armstrong crawled through a tiny hatch into the Eagle and

184

began the extensive preparations necessary to descend to the lunar surface. Around 2 p.m., Armstrong and Aldrin detached the Eagle from the Command Module and both spacecraft began to orbit the Moon.

As the spacecraft disappeared behind the Moon, Kranz seized the opportunity to deliver a pep talk to the mission controllers. Using a private communication loop to ensure that his message only went to his team, he reminded everyone that the entire world would be watching them and that they were about to make history and attempt to do something that had never been done before. Kranz explained that he had complete confidence in every one of the controllers, and that whatever happened, he would stand behind their every decision.

That moment has remained firmly in Bales's mind to this day, and he can still remember Kranz's final remark:

> He ended by saying that no matter how it goes, we would walk out of that room as a team. Those words had an incredible impact on me. He was telling us that we were as prepared as we could be, and that he would support our decisions and that we were a team. If it didn't end in success, it was not going to be about blaming any one person. We walked into that room as a team and we would walk out as a team. It really fired me up, and also took some of the stress off. It was just what I needed to hear.[144]

Kranz then asked for the doors of Mission Control to be locked. He didn't want his controllers to be distracted by people coming in or going from the room, and perhaps more importantly, wanted to remind the team that they were now fully responsible for what was about to happen.

Right on cue, both the Command Module and the Eagle re-appeared on Mission Control's screens, and Armstrong and Aldrin began their journey to the lunar surface. As they sped towards the Moon, communications proved unreliable and they struggled to speak with Mission Control. Worse still, they seemed to be off course. Unbeknownst to anyone, when the Eagle had undocked from the Command Module, the air in the tunnel connecting the two spacecraft hadn't been fully removed. As a result, the small amount of air remaining had given the Eagle a tiny extra push and it was now heading outside its planned landing zone.

Doug Ward was providing live commentary on the landing and can remember the tension in Mission Control:

'All the way down the communication with the astronauts was patchy. We kept losing it and getting it back. I was on the edge of my seat and kept thinking that we were going to have to abort.'[145]

The situation went from bad to worse. As they approached the Moon, Mission Control heard Neil Armstrong report: 'Program alarm.' The module's onboard guidance computer had flashed up error code '1202'. Concerned, Armstrong radioed back to Mission Control: 'Give us a reading on the 1202 program alarm.'

Back on the Earth, it now fell to Bales to decide whether to abort the landing and there was no time for lengthy debate. Bales's response had its roots in a form of thinking that was vital to the success of the entire Apollo programme and we will return to his dilemma in a moment.

THINKING 'WHAT IF . . .'

In the opening chapter, we met Jerry Woodfill. On a basketball scholarship from Rice University, Woodfill was not doing especially well in his studies. When Kennedy came to the Rice University Stadium, Woodfill went along and heard the President give his famous speech about putting a man on the Moon before the decade was out. Inspired by Kennedy's words, Woodfill dropped his basketball and started to study electrical engineering. After graduation he applied to NASA and was asked to help design the safety systems for the Lunar Module.

Occupying part of the building next to Mission Control (known as the Mission Evaluation Room), Woodfill and his colleagues provided technical assistance to the astronauts and flight controllers. Part of the work involved imagining the tricky scenarios that might emerge during a mission and then trying to come up with ways to avoid the problem or cope with the issue. At one point, Woodfill was thinking about the radar system that the astronauts would use to land on the Moon and an idea suddenly popped into his mind. The radar contained an alarm that would sound if it started to overheat. After the Lunar Module touched down on the surface of the Moon, the radar was no longer important, but the overheating alarm would still be operational. What would happen, wondered Woodfill, if the astronauts set down on the Moon and then the heat from the Module's engine accidentally activated the radar's temperature alarm? Woodfill worried that if this happened while the astronauts were out exploring the Moon, they might be forced to return to the Module to figure out what was going on and so unnecessarily cut short their historic walkabout.

Curious, Woodfill and his colleagues ran a thermal analysis on the radar system and discovered that the Module's engine heat might indeed trigger a false alarm. It cost very little to fix the potential problem and Woodfill estimated that this example of 'what if . . .' thinking had saved the Apollo programme millions of dollars and a huge amount of potential embarrassment.

'What if . . .?' took place at every level of the Apollo programme. In the White House, Richard Nixon's advisors had prepared a speech for the President to deliver in the event of the Apollo 11 Moon landing ending in catastrophe.[146] The haunting words mourn the lost astronauts and humanity's quest for exploration:

Fate has ordained that the men who went to the Moon to explore in peace will stay on the moon to rest in peace. These brave men, Neil Armstrong and Edwin Aldrin, know that there is no hope for their recovery. But they also know that there is hope for mankind in their sacrifice . . . For every human being who looks up at the Moon in the nights to come will know that there is some corner of another world that is forever mankind.

Thankfully, Nixon was never called upon to deliver the speech, and it was hidden away in the National Archives and only made public thirty years after the mission.

By far the most extensive and complex type of 'what if . . .' thinking took place in Mission Control. One wall of Mission Control contained several large windows, and the area behind these windows was home to the 'sim' (short for 'simulation') teams. This group of around thirty engineers and scientists would carefully examine the Apollo planning documents and imagine what might go wrong on a mission. They would then run a simulation session in Mission Control and see if the controllers could cope with the problems. The controllers knew the purpose of the simulated mission, but had no idea what problems were about to come their way. The work was complex and relentless, with the simulations sometimes running several times a day and six days a week.

Electrical engineer Jay Honeycutt played a key role in the Apollo sim team. Much of his work involved feeding data into Mission Control to simulate specific issues. Would the controllers notice if fuel started to leak out of the spacecraft? How would they react if an engine malfunctioned? Could they cope if the Lunar Lander suddenly shot off course? On some occasions the simulations became less data-driven and far more physical. Before one exercise, Honeycutt secretly tied a piece of string to the power input on a controller's console and ran the string under the floor and out into the sim room. During the simulation, Honeycutt waited until the controller was about to make a critical decision, and then yanked on the string and cut off the power to the console. Honeycutt was delighted to see the controller quickly switch consoles and carry on with the mock mission.

During another sim, Honeycutt and his team decided to trip the circuit breakers that supplied the power to Mission Control. About a third of the consoles and half the lights suddenly went down. The controllers took a long time to figure out which of the breakers had failed and find replacements. However, as Honeycutt recalls, Mission Control was quick to rectify the issue:

I arrived at 6.30 the next morning and there were these huge drawings spread out all over the floor. The guys were on their hands and knees colour-coding each of the wires. Within a week there was a number on every breaker, and they could locate and replace the right one in five minutes. That was very much the attitude. Nobody ever got mad at us. We were in this together and this was the best way to learn.[147]

This work of the sim teams was vital to the success of the Apollo programme. As Gerry Griffin commented:

When it came to our missions, around 90% of our time was spent trying to figure out what we are going to do if this happens, and what we are going to do if that happens. We had backup systems or procedures for everything that we could. And there's no question about it, all the preparation saved us. I mean really saved us many, many times because I don't think there was a single mission that we didn't have some significant problem.[148]

Just a few weeks before the Apollo 11 landing, Kranz's team appeared overconfident and so simulation supervisor ('SimSup' for short) Dick Koos decided to make life more difficult for them. Koos had been associated with America's space-flight programme from its infancy and was one of the most respected authorities in the simulation of space-flight missions. He instructed his team to load 'Case Number 26' into the system to discover whether Kranz's team could deal with the alarms produced by the Lunar Module's onboard computer.

During the simulation, the Lunar Module's Guidance Computer suddenly delivered a '1210' alarm.[149] Although the computer that would help guide the astronauts to the lunar surface was state of the art in the late 1960s, it had less computing power than a smartphone. The '1210' alarm indicated that two pieces of hardware were attempting to communicate with the computer at the same time and that the computer was struggling to cope.

Steve Bales was working the console that day, and when the alarm popped up Bales was uncertain what was going on. He could see that the computer was struggling but wasn't certain whether that was mission critical and so decided to err on the side of caution and abort the landing. During the debriefing session, the mission controllers discovered that the code wasn't serious enough to justify an abort and that they should have continued the landing.

Koos ticked off Bales and Kranz for an unnecessary abort and Kranz's team was put through several hours of training on programme alarms. The software developers assured Mission Control that the type of alarm that had come up during the simulation had been created for troubleshooting the computer code before it was released, and that these sorts of alarms would be very unlikely to occur during an actual mission. Nevertheless, Kranz asked his team to take a look at all the alarms that could conceivably come up.

Bales asked his close colleague, software support engineer Jack Garman, to review all possible alarms and to identify those that were critical to the mission. A few days later, Garman showed Bales a handwritten crib sheet containing a summary of his findings and

Bales reviewed each of the alarms and signed off the sheet. Little did the two of them know that their work would prove vital when it came to the historic Apollo 11 landing.

Moonshot Memo
Subject: Tindallgrams

Bill Tindall Jr was an exceptional engineer, a talented manager and a witty man.[150] Tindall helped build America's space programme from the bottom up and had a remarkable ability to motivate those around him, think on his feet, get his head around complex issues and coordinate diverse groups of people. Flight Director Gene Kranz once described him as the 'architect for all of the techniques that we used to go down to the Moon',[151] and Kranz invited Tindall to sit beside him in Mission Control when Neil Armstrong piloted the Lunar Module down onto the surface of the Moon.

Early in his career, Tindall started to send memos to his colleagues to help focus their attention on key issues.

Over the years, Tindall penned more than a thousand memos. Although they tackled serious issues, many of the memos had light-hearted titles ('Vent bent descent, lament'; 'Happiness is having plenty of hydrogen') and contained humorous asides ('If the data is right, we are in deep trouble with a capital "S"). As a result, the memos were widely circulated and eventually became affectionately known as 'Tindallgrams'.

In one memo, Tindall pointed out that immediately after the Lunar Lander had touched down on the Moon, the flight controllers would have just a few moments to decide whether the astronauts should remain on the lunar surface, or abort and try to make their way back up to their orbiting spaceship. At the time, the word 'Go' was being used to indicate that an operation could proceed and 'No Go' to indicate it should be aborted. Tindall had spotted that this could prove confusing after the landing. Did 'Go' mean stay and 'No Go' mean abort? Or did 'Go' mean abort and 'No Go' mean stay? Mission Control solved the problem by following Tindall's advice, and the terminology was altered to 'Stay' and 'No Stay'.

A simple misunderstanding could have ruined one of humanity's greatest moments. Instead, Tindall's clear thinking and straightforward way of communicating key issues helped save the day.

THE POWER OF PREPAREDNESS

On 20 July 1969, the Eagle was descending rapidly to the Moon and the onboard computer had just signalled a '1202' programme alert. A potentially life-or-death struggle was taking place 350,000 kilometres from the Earth.

This time, however, the team was prepared. Garman quickly consulted his crib sheet and identified programme alarm '1202' ('Executive overflow'). The Eagle's navigational computer had a lot on its mind. It was trying to calculate the distance to the surface of the Moon, relay important information back to Mission Control and keep an eye on the position of the Command Module in case Armstrong had to abort the landing and head back up. In addition, and unknown to Mission Control and the astronauts, a tiny glitch in the system meant that the computer was being asked to process extra information that wasn't required for the landing.

Unfortunately, it had all become a bit too much and the computer had completed as many of the tasks as possible, saved what it could and started to reboot.

Garman realized that it was fine to continue with the landing as long as the code didn't permanently appear on the screen and instantly communicated his findings to Bales. Bales had to combine the information from Garman with the navigational data coming in from the Lunar Module, and quickly decide whether to abort the landing. With the eyes of the world on him, twenty-six-year-old Bales stepped up to the mark and made the call to continue. Hundreds of thousands of miles away, Armstrong and Aldrin carried on with their descent.

CAPCOM Charlie Duke relayed the good news: 'Roger. We're GO on that alarm.'[152]

A few moments later, Aldrin relayed another problem: 'Roger. Understand. GO for landing. 3000 feet. PROGRAM ALARM.'

This time it was a '1201' alarm. Once again Garman checked the list ('Executive overflow – no vacant areas') and Bales gave the landing the green light.

CAPCOM: 'Roger. 1201 alarm. We're GO. Same type. We're GO.'

As the Eagle approached the Moon a second major issue emerged. The Apollo team had spent years studying photographs of the lunar surface to identify an ideal landing site. They had decided on an eleven-mile long by three-mile wide elliptical area known as 'Mare Tranquillitatis' ('Sea of Tranquillity'). Unfortunately, the extra push that the Eagle had received when it separated from the Command Module had resulted in it overshooting its original landing site. Worse still, it was now heading directly towards a vast crater dotted with huge car-sized boulders – colliding with any of these giant rocks would almost certainly destroy the Eagle. The endless simulations were about to pay off. Armstrong calmly took over manual control of the craft and started to skim over the boulders. Less than a hundred feet above the lunar surface, the Eagle had very little of its fuel left and could only remain airborne for another minute. Mission Control fell silent and CAPCOM relayed the information:

CAPCOM: 'Sixty seconds.'

Dropping below forty feet, the situation became even more problematic when the Eagle's engines began to kick up the Moon's dust and severely reduce visibility.

CAPCOM: 'Thirty seconds.'

With only twenty seconds of fuel remaining, Armstrong finally found a suitable landing spot and the Eagle gently touched down on the surface of the Moon. The Eagle was around four miles away from the original landing site.

Aldrin: 'Contact Light. Okay. Engine Stop.'

The Eagle's engines shut down. Outside, the Moon dust that had sat undisturbed for over a billion years started to settle. Armstrong then uttered the words that the whole world had hoped to hear:

Armstrong: 'Houston, uh . . . Tranquility Base here, the Eagle has landed.'

Excited, but somewhat tongue-tied, Charlie Duke replied:

CAPCOM: 'Roger, Twank . . . Tranquility, we copy you on the ground. You got a bunch of guys about to turn blue here. We're breathing again. Thanks a lot!'

Doug Ward remembers the sense of relief in the room:

I heard an enormous cheer from the politicians, managers and astronaut families in the viewing room behind us. Kranz quickly spoke to all of the mission controllers and had everyone settle down, and then started to work through the stay/no stay checklist. Everything was fine. We were on the Moon.[153]

Millions of people across the world saw the seemingly impossible happen right before their eyes. That night, someone placed a small bouquet of flowers and a handwritten note next to the eternal flame on President Kennedy's grave at Arlington National Cemetery. The card simply read: 'Mr President, the Eagle has landed.'

Some of the mission controllers took a well-earned rest. Flight Controller Ed Fendell went home, had some sleep, and then headed back to work. On his way he dropped into to a local diner and grabbed some breakfast:

I ordered scrambled eggs. Two guys came in and sat down on the stools next to me. They worked at a local gas station and they started to chat about the landings. One of them said: 'You know, I went all through World War Two. I landed at Normandy on D-Day. It was an incredible day. And then I went through Paris and on to Berlin. But yesterday was the day that I felt the proudest to be an American.'[154]

Ed had been working day and night in Mission Control, and hadn't realized the impact that the Moon landing had had on his fellow countrymen. Ed felt himself choking up and paid for his eggs. He then picked up his newspaper, went to his car and started to cry.

Several months after the safe return of the Apollo 11 crew, President Richard Nixon hosted a grand dinner in Los Angeles to honour those involved in the historic mission. In front of an audience of leading politicians and ambassadors, Nixon posthumously awarded the Distinguished Service Medal to the three astronauts who had lost their lives in the Apollo 1 fire, the Presidential Medal of Freedom to the Apollo 11 astronauts and the NASA Group Achievement Award to the mission operations team. Steve Bales had the honour of accepting this final award. As he stood proudly on the podium, Nixon declared: 'This is the young man, who when the computers seemed to be confused and when he could have said "Stop" said, "Go."' And all because of the power of preparedness.

*

Being Prepared

'By failing to prepare, you are preparing to fail.'
– Benjamin Franklin

Let's start with a quick questionnaire.

Please think of a time when you were anxious before an event. Perhaps you had to give a talk to your work colleagues, attend a job interview or go to a party where you didn't know anyone. Next, please read the following ten statements and rate the degree to which each one describes your thoughts prior to the event. Assign each statement a score between 1 ('I never thought something like that') and 5 ('That's exactly what I thought').

		Your score
1	In general, I worried about how everything would turn out.	
2	I sometimes thought about specific things that might go wrong.	
3	I had a feeling that I wouldn't be able to cope at the event.	
4	I tended to spend quite a bit of time thinking about the problems that might emerge at the event.	
5	I felt as if the outcome of the event would be outside my control.	
6	In general, I imagined how I could solve any problems that might emerge.	
7	I tried not to think about the event until the last minute.	
8	I thought that imagining what could go wrong during the event would help me to prepare.	
9	I tried not to think about the event because that made me feel anxious.	
10	I often thought about the worst that could happen, even though deep down I suspected that this would never actually happen.	

We will return to the questionnaire in a few moments.

Psychologists have conducted thousands of scientific studies examining the impact that optimism and pessimism have on people's lives.[155] Over the years, several clear patterns have emerged. In general, optimists are physically healthier and psychologically happier than their more pessimistic counterparts. Also, as we discovered in Chapter 3, optimists are far more likely to start to achieve their goals and keep going when the going gets tough. All this adds up over the long haul, resulting in those looking on the bright side of life tending to be especially successful in both their personal and professional lives. Over time, these findings have made their way into the public domain, with musicians, writers, therapists and self-help gurus all promoting the advantages of walking on the sunny side of the street.

Then, a few years ago, that all started to change.

Psychologist Julie Norem, from Wellesley College in Massachusetts, has spent much of her career taking a more fine-grained look at pessimism and has discovered not all pessimists are created equal.[156] In general, most pessimists expect the worst, are fatalistic about life and aren't much fun to be around. However, Norem's work has revealed that roughly a third of pessimists have a rather unusual thinking style known as 'defensive pessimism'. Defensive pessimism puts a positive spin on negative thinking. Whereas most pessimists run away from whatever makes them feel anxious, defensive pessimists identify problems that might crop up, and take steps either to prevent these issues or to think of ways of dealing with them.

Let's take a closer look at defensive pessimism in action. Imagine that you are going to an important job interview. If you are an optimist, then you will probably expect the interview to go well. Similarly, if you happen to be a run-of-the-mill pessimist, then you will tend to have a general feeling that things will go badly. However, if you are a defensive pessimist, then you will think about the ways in which the interview might be a disaster and the steps that you can take to prevent these problems. Maybe you worry about being late for the interview and so plan to arrive especially early.

Maybe you are concerned about being asked a tricky question and so spend some time thinking about possible answers to all your nightmare questions. Or perhaps you are worried that you will appear overly nervous and so rehearse the interview with your friends and colleagues. In short, defensive pessimists embody the type of 'what if . . .?' thinking that helped humanity get to the Moon.

Please take a look at the questionnaire you completed a few moments ago, and add up your scores to all the ODD-numbered statements (1, 3, 5, 7 and 9). This reflects your general optimism–pessimism score, with a low score (between 5 and 10) indicating that you always look on the bright side of life and a high score (between 20 and 25) suggesting that you have a more gloom-oriented worldview.

Next, add up your scores to the EVEN-numbered statements (2, 4, 6, 8 and 10). These statements are the type used by psychologist Julie Norem in her research and reflect your level of defensive pessimism. A low score (between 5 and 10) suggests that you don't tend to engage in contingency planning, whereas a high score (between 20 and 25) suggests that you like to identify possible problems and think about how best to deal with every eventuality.

Fortunately, there are several techniques to help you perform at your best by preparing for the worst. It's just a case of visiting the theatre of your mind, conducting a pre-mortem, and meeting the mythical daughter of the King of Troy.

THE THEATRE IN YOUR MIND

There are several techniques that can be used to help generate 'what if . . .?' scenarios and then think through possible solutions to potential problems. Perhaps the most effective approach involves doing what the Apollo team did, namely, rehearsing. Sometimes it's possible to run mini-simulations to expose potential pitfalls. For instance, if you are thinking of writing a book, try producing a blog post each day for a month. Or if you are considering launching a new product, try testing the market by producing a small-scale prototype. Or if you are weighing up the pros and cons of becoming self-employed, try working for yourself at the weekends and see how it goes.

Unfortunately, sometimes it isn't possible or practical to rehearse. However, the good news is that you can use a technique that many athletes, musicians, actors, CEOs and entrepreneurs employ to great effect: mental rehearsal.

To illustrate how the technique works, let's imagine that you have to give an important slide-based presentation at work.

1) First, grab a pen and paper, and find a place where you won't be disturbed. Draw a vertical line down the middle of the paper and write the word 'PROBLEM' at the top of the left-hand column.

2) Next, it's time for a spot of mental rehearsal. You are about to spend a few moments thinking about how the presentation will go. You might be tempted to imagine everything going really well, and to see the whole room clapping and

cheering at the end of your talk. Resist the temptation. The goal isn't to inflate your ego, raise your hopes and live in a make-believe world were the sun always shines. Instead, it's all about being realistic and preparing for possible failure.

Please come up with ten potential problems with the presentation. Away you go.

	PROBLEM	
1		
2		
3		
4		
5		
6		
7		
8		
9		
10		

How did you do? Typical issues include:

	PROBLEM	
1	The computer won't connect to the projector and so the audience can't see my slides.	
2	The microphone doesn't work and so the audience can't hear me.	

3	I might be overly nervous at the start of a talk.	
4	I could stumble over my words.	
5	My mind might suddenly go blank.	
6	I might move a slide on too early.	
7	I will lose my place in the presentation.	
8	One of my jokes will fall flat.	
9	I might under-run and won't be able to fill the time.	
10	I might over-run and people have to leave before I have finished.	

3) Now that you have identified several potential problems, it's time to figure out either how to prevent the issue in the first place, or how to cope if it does occur. Please write the word 'SOLUTION' at the top of the right-hand column. Take a look at each of the items that you have listed in the 'PROBLEM' column and think of how best to prevent, or cope with, the issue. Jot each of these ideas down in the 'SOLUTION' column. Feel free to list as many solutions as possible for each of the problems.

How did you do? Typical issues include:

	PROBLEM	SOLUTION
1	The computer won't connect to the projector and so the audience can't see my slides.	Can you go to the room beforehand and check the equipment? Could you bring your own small projector, and use that in an emergency? Is there a version of the talk that you could give without the slides?
2	The microphone doesn't work and so the audience can't hear me.	Could you cope by simply speaking up? Is there a simple version of the talk and can you get the gist of the talk from the slides alone?
3	I might be overly nervous at the start of a talk.	Is it helpful to tell yourself that you are excited rather than nervous? Or carry out a quick relaxation exercise? Or have your opening line in mind?
4	I could stumble over my words.	Could you just acknowledge the stumble and keep going? Or have a line ready, such as 'I know what you are thinking: That's not an easy thing to say'?
5	My mind might suddenly go blank.	Can you apologize, take a drink of water to buy some time, back-up a slide and move on? Can you have a card with the key points on? Can you have some kind of line ready, such as 'I have seen this happen to other speakers and wondered how it felt: now I know'?
6	I might move a slide on too early.	Can you make a little joke, such as 'Let's all imagine that you didn't see that yet'?

7	I will lose my place in the presentation.	Could you back up one slide? Or reach for your emergency notes? Or admit what has happened and ask the audience where you were in the presentation?
8	One of my jokes will fall flat.	Can you have a couple of lines ready, such as 'That joke normally plays better than that, but I think you are right' or 'And that's the last time I will be doing that joke.'
9	I might under-run and won't be able to fill the time.	Can you ask for questions? If you don't get any, how about saying something like 'One of the most frequent questions is . . .', so that you have something to talk about?
10	I might over-run and people have to leave before I have finished.	Can you keep an eye on your watch or a clock? Or set a timer on your phone to sound five minutes before the end? Do you have a final slide that you can cut to if you run out of time?

By asking 'what if . . .?' you have identified the main issues, and thought about how best to prevent or deal with them. Just like the Apollo 11 sims team, you will have sown the seeds of success by preparing for the worst.

Thinking like a defensive pessimist will not only help you prepare, but will also help banish general feelings of anxiety. The next time that you start to feel generally worried about a situation, identify specific problems and, more importantly, think about what you can do to help alleviate these potential issues.

CONDUCT A PRE-MORTEM

Carrying out either an actual rehearsal, or a mental rehearsal, is a great way of contacting your inner defensive pessimist. However, it's not the only approach. Another effective technique involves conducting a 'pre-mortem'. This two-part technique was invented by psychologist and expert on decision-making Gary Klein.[157] During a post-mortem you take a look at someone who has passed away and try to figure out why they died. During a pre-mortem you do the same with any project or enterprise, but carry out the procedure before you have begun. Here's how to do it.

First, you have to engage in a spot of mental time travelling. Imagine yourself in the future and that your venture has been a spectacular failure. Ask yourself one question: Why did it all go so badly wrong? No idea is off-limits. If you are conducting the pre-mortem with a group, everyone should feel completely uninhibited about mentioning any problem, no matter how ridiculous it sounds. If you are working on your own, be brutally honest. Maybe no one shows up to your event, the project website crashed, a key person backed out at the last minute. In short, you assume that the 'patient' has died and that everything went as badly as possible. It's your job to figure out why it was all so disastrous. Klein has designed this part of the technique to stop people worrying about seeming overly negative by actively inviting them to contact their inner devil's advocate.

Second, select the top ten problems that have emerged, and try to find solutions. If you have managed to think of more than ten problems, try to focus on the 'show-stoppers' that are mission critical.

The next time that you want a project to come alive, spend a few moments assuming that it died a terrible death.

THE CASSANDRA COMPLEX AND OTHER PROBLEMS

According to Greek mythology, the King of Troy had a beautiful daughter called Cassandra. One day, the God Apollo took a shine to Cassandra and decided to try to court the love of his life by presenting her with a gift. Many mere mortals would have gone for a bunch or flowers or a box of chocolates. However, Apollo went big and removed the element of surprise from Cassandra's life by granting her the gift of prophecy. Unfortunately, Cassandra was less than impressed with Apollo and refused his magical advances. Furious, Apollo got even by placing a curse on Cassandra, using his omnipotence to ensure that no one would ever believe her predictions. For the rest of her life Cassandra was able to foresee the future but wasn't able to convince people that her predictions would prove accurate.

In one incident, Cassandra told the good folks of Troy that their city would be attacked by a group of Greek soldiers hiding in a giant wooden horse. Apollo's curse ensured that Cassandra's comments fell on deaf ears and her predictions were ignored. The Greeks then declared that they were ready to end the long-running Trojan War and presented the city of Troy with a giant wooden horse as an act of goodwill. When Cassandra saw the horse, she grabbed a burning torch and attempted to set fire to it. Still in denial about the accuracy of her visions, the Trojan people wrestled Cassandra to the ground and brought the horse into their city. Shortly afterwards, a group of Greek soldiers sneaked out of the horse and destroyed Troy.

There's much to be learned from Cassandra's story. First, if you aren't interested in the romantic advances of a Greek God, find a way of letting them down gently. Second, and perhaps more importantly, people often don't like to hear predictions of doom and gloom. This phenomenon, known as the Cassandra Complex, can cause problems for defensive pessimists. Most people don't enjoy hearing about everything that might go wrong and the need for contingency plans. In addition, defensive pessimists may sound overly anxious and even incompetent. In some scenarios it might be wise to consider keeping some of your more pessimistic thoughts to yourself, or minimize the chances of getting a reputation for being too negative by openly acknowledging that you are super-cautious, are playing devil's advocate, are worrying about the situation because you care about it or are trying to do something that will help everyone in the long run.

This is not the only issue associated with defensive pessimism. When you start to think about potential problems, your thoughts may run the risk of becoming quite extreme. For instance, when it comes to giving a presentation you might imagine being attacked by the slide projector or the ceiling caving in. Similarly, your thoughts might start to spiral out of control. You might, for instance, think something along the lines of: 'I could fluff my lines and then the audience will laugh at me. If that happens, my boss will think that I am a fool. I am already in her bad books and this might be the final straw. If she makes me redundant then I will have no way of paying the rent, and so I will have to go back to living with my parents. That would be a disaster.'

Avoid both issues by not wasting time thinking about problems that are very unlikely to happen, and focusing on solutions rather than how one problem will lead to another.

Finally, once in a while you might come across a problem for which there isn't any obvious solution and this might make you feel especially worried. If there really is something that you can't control, don't waste time worrying about it because, by definition, there's nothing you can do. If you need some inspiration, think back to the last chapter and the way in which Flight Director Glynn Lunney handled the first time that Mission Control lost contact with Apollo 8. The three astronauts went behind the Moon and had to carry out a tricky engine burn. There was nothing that the controllers could do. Rather than have people sitting there worrying, Lunney told people that now was a good time to take a comfort break. The next time you are faced with a worrying situation that is outside your control, think like Lunney and his team, and find something to take your mind off the situation.

SUMMARY

In general, pessimism isn't good for you. However, defensive pessimism is helpful because it encourages you to think 'what if . . .?', and then come up with helpful contingency plans. To encourage this type of thinking:

– Find a way of creating your own Apollo 'sims' team to help you cope with the various eventualities that may arise. This might take the form of an actual rehearsal or a mental rehearsal. In addition to making you fully prepared, it will help prevent a general sense of anxiety.

– Carry out a pre-mortem. Imagine that your project has already failed. What went wrong and why? What can you do to prevent these problems?

– Defensive pessimism is a powerful way of thinking, but don't be too extreme, find a way of gently raising your worries, don't let your thoughts spiral out of control, and don't waste time worrying about events that you can't control.

8

BUZZ ALDRIN AND THE MISSING SWITCH

Where we discover how astronauts had to improvise
their way home, and learn how you can thrive in the
face of the unexpected.

The plan had been for Armstrong and Aldrin to have a good sleep and then step out onto the surface of the Moon. Perhaps not surprisingly, the two astronauts were excited about exploring their new surroundings and so decided to skimp on sleep and venture out as soon as possible.

They were about to encounter an extremely hostile environment. The Moon has hardly any atmosphere and, as a result, there's little to deflect or absorb the energy from the Sun. During the day, the Sun's rays can heat the surface of the Moon to temperatures in excess of 100 degrees Centigrade. At night, or in shadow, the Moon becomes one of the coldest places in the solar system, with temperatures dropping to a terrifying minus 170 degrees Centigrade.

These extreme temperatures were just the tip of the iceberg. The two astronauts would also face an onslaught of micrometeoroids

– tiny particles of rock flying through space at terrifying speeds. These particles packed a punch way above their weight, and the grey powder on the lunar surface was the result of vast numbers of them pulverizing rocks into dust. Worse still, the Moon has no oxygen, no air pressure, and is constantly subjected to several forms of extremely dangerous radiation.

As a result, preparing for a Moonwalk involves far more than putting on a warm coat and a sensible pair of shoes. In fact, it was another dramatic example of extreme preparedness. Armstrong and Aldrin's safety depended on one of the most elaborate and expensive outfits ever created. Each astronaut's spacesuit cost an estimated $100,000 (around $700,000 today) and if any part of it failed the effects would quickly be fatal.[158]

The astronauts' underwear consisted of a tight-fitting onesie containing hundreds of feet of thin tubing. During their time on the lunar surface, chilled water was continuously circulated through the tubing to prevent the astronauts from overheating.

The spacesuits themselves were created by the International Latex Corporation (best known for producing Playtex bras and girdles), and consisted of twenty-one layers of high-tech materials. Each suit was custom-made by highly skilled seamstresses and required incredible precision. The stitches on some of the seams were less than a millimetre long and the slightest error could mean the difference between life and death. When the astronauts ventured out of their Lunar Module, their spacesuits were inflated to create a breathable and pressurized environment and so it was vital that the multi-layered outfit didn't have any leaks or weak spots. As astronaut Jim Lovell memorably remarked to one seamstress: 'I would hate to have a tear in my pants while on the Moon.' Learning lessons from the tragic Apollo 1 fire, the spacesuit's outermost layer consisted of a Teflon-coated cloth that could withstand over 1,000 degrees Fahrenheit, and also protected the astronauts from dangerous solar radiation and the dreaded micrometeorites.

The astronauts' gloves were complex and contained wire cables to ensure structural support and flexibility in a vacuum, silicone fingertips to allow the astronauts to manipulate objects and steel fibre cloth to prevent them being cut by rocks or tools. The gloves were securely connected to the arms of the spacesuit via large metal rings; heavy overshoes ensured that the astronauts kept their feet firmly on the ground at all times.

Finally, each astronaut had to wear a large backpack that contained their oxygen supplies, maintained the pressure in the spacesuit and kept the chilled water circulating through their undergarments. All being well, this self-contained life-support system could keep them going for a maximum of four hours.

Moonshot Memo
Subject: Apollo, pinball and the Orbitor 1

The astronauts' gloves were designed by an engineer and inventor named Dixie Rinehart. Like so many of those working on the Apollo programme, Dixie spent his childhood playing with rockets and coming up with new inventions. It wasn't all child's play – on one occasion his rocket-based endeavours blew out the windows of his parents' house and created a large crack in their cellar wall.[159]

Dixie ended up working for the International Latex Corporation and spent years designing the gloves that would help keep Armstrong and Aldrin safe on their historic Moonwalk. He is one of only eight people to be listed on the official patent for the Apollo spacesuits.

After Apollo, Dixie continued to develop his glove-based technology and eventually created a unique form of work glove that turned out to be very handy and sold by the millions. Dixie also worked on other inventions and, in the later 1970s, he and his colleagues created a pinball game known as the 'Orbitor 1'.[160] It is the first and only pinball machine to have a contoured playfield rather than a flat surface. This weird playing field was inspired by Einstein's ideas about space–time distortion and had the appearance of a lunar landscape. The idea came to Dixie after Apollo astronaut Wally Schirra had arranged for him to give a presentation at Walt Disney World. Some reviewers remarked that the Orbitor 1 was unlike any pinball game they had ever seen. As one of Dixie's colleagues remarked, that was mainly due to the inventors knowing nothing about pinball.

Dixie's enthusiasm, passion, imagination and enormous sense of fun was vital to his work. As we have seen time and again, these traits played a surprisingly important role in the success of Apollo.

The Greatest Show Off Earth

A few hours after landing on the Moon, the two astronauts were suited up and moments away from ensuring their place in history. Hundreds of thousands of kilometres away, the world watched in wonder, with the weekly entertainment magazine *Variety* describing it as 'The Greatest Show Off Earth'.

In America, CBS television staged a thirty-one-hour 'super-special' that was presented by legendary news-anchor Walter

Cronkite. Similarly, the ABC network broadcast their own marathon programme, filling the time with several space-related items, including Twilight Zone creator Rod Serling hosting a panel of famous science-fiction writers and Duke Ellington performing a specially commissioned composition, 'Moon Maiden'. The coverage was broadcast live to huge screens across the nation. Central Park's Sheep Meadow was renamed 'Moon Meadow' for the occasion and everyone going along to watch the public viewings was urged to dress in white.

In Britain, the ITV channel broadcast sixteen hours of continuous coverage in the form of David Frost's *Moon Party*. A somewhat strange mix of news and entertainment, the show interspersed information about the landings with a variety of more light-hearted interludes, including songwriter Cilla Black singing her latest hit and comedian Eric Sykes performing a sketch about a bullfighter from Manchester.[161] Other highlights included a caller from Eastbourne asking experts whether it would be possible to use Moon dust to grow especially large pumpkins, and a joint discussion between historian A. J. P. Taylor and entertainer Sammy Davis Jr about the ethics of manned space flight.

On the BBC, it was a slightly more sober affair, with the coverage featuring Pink Floyd playing an improvised piece entitled 'Moonhead', famous actors reading poetry about the Moon and footage from the Apollo mission being juxtaposed with the David Bowie song 'Space Oddity'. (Ironically, many music critics believe that Bowie had intended his song to be an antidote to what he perceived as 'space fever' and that the lyrics tell of an astronaut who is both physically and psychologically lost in space). Similar television programmes across the globe ensured an audience in excess of 500 million people, making the landings one of the most watched events of all time. The whole world had become moonstruck.

On 21 July 1969, Armstrong wriggled out of the Lunar Module's hatch and made his way down the spacecraft's ladder. In order to make the Module as light as possible, the ladder was built from aluminium and was so flimsy that it would have struggled to support Armstrong's weight back on the Earth. As he neared the bottom of the ladder, Armstrong switched on the Module's outside television camera and the low-resolution, black-and-white images were transmitted back to the Earth.

The BBC coverage was co-presented by well-known science journalist James Burke and the historic broadcast is still fresh in his mind:

> It happened in the middle of the night for Britain, and half the country had stayed awake to watch. There was an enormous pressure to get everything right. At first, there was nothing to see. When Armstrong went down the steps to put his foot on the Moon, there was no camera covering him until the very final moment. I had directors saying into my ear, 'Describe what we're looking at'. Nobody had ever seen something like that before, so it was tricky to describe![162]

The world watched in awe as a white blob slowly moved down to the bottom of the ladder, and then tentatively stepped onto the dusty surface below. For the first time in history, a human had set foot on the Moon. Moments later, Armstrong uttered his now famous words: 'That's one small step for man, one giant leap for mankind.' Armstrong always maintained that static noise deleted an all-important 'a' and that in reality he declared: 'That's one small step for *a* man, one giant leap for mankind.' Either way, it was no small step for Armstrong. The touchdown had been so gentle that the Lunar Module's shock absorbers hadn't fully compressed and so Armstrong's small step was actually closer to a four-foot jump.

Once again, James Burke remembers the moment well:

It was a difficult moment as a broadcaster. The worst thing I could do was talk over the top of an astronaut, but I never knew when they were going to speak. So my brain was split into two, with one part trying to keep talking and the other part being ready to shut up in a second if I had to. I was terrified about speaking over Armstrong when he said, 'That's one small step for man', and so I was far more concerned with not messing up than enjoying the extraordinary thing that was happening. And then afterwards it occurred to us that it had been a spectacular event.[163]

THE BEST-LAID PLANS . . .

Armstrong reported that the Moon dust was only a few inches deep and collected some rock samples. Around twenty minutes later, he was joined by Aldrin and the two astronauts began their Moonwalk. The astronauts started off by placing a commemorative stainless steel plaque on the lunar surface ('*Here Men From Planet Earth First Set Foot Upon The Moon. July 1969 A.D. We Came In Peace For All Mankind.*').

They then hastily erected an American flag, complete with a hidden horizontal crossbar to give the impression that the flag was flapping in the wind. Next, the two astronauts took the ultimate long-distance telephone call and had a brief conversation with President Nixon (Nixon: 'As you talk to us from the Sea of Tranquillity, it inspires us to redouble our efforts to bring peace and tranquillity on Earth'). Finally, they bounced around to test their mobility and gathered some more samples of dust and rocks.

All this activity was fed back to the Earth via a live television feed and recorded for posterity with a specially adapted stills camera. Armstrong carried the stills camera much of the time and so took the majority of the hundred or so photographs. As a result, the photographs contain around twenty striking images of Aldrin carrying out various tasks, but not a single similarly impressive image of Armstrong. Perhaps the most iconic shot of Aldrin shows him standing on the lunar surface looking vulnerable and fragile. Over the years, the image has been reproduced countless times in books, magazines and websites. Look closely and there, clearly reflected in Aldrin's gold visor, you will see Armstrong. The first man on the Moon doesn't feature in the photographs from the historic trip, but on the upside, inadvertently, he produced the ultimate 'selfie'.

The astronauts spent just over two hours on the Moon, and then returned to the Lunar Module.

The Apollo 11 mission seemed to be proceeding like clockwork. However, a problem was lurking in the shadows. When Armstrong and Aldrin had been moving around in the Lunar Module, one of them had accidentally knocked into the wall of their spaceship and broken off the outside portion of one of the switches.

Unfortunately, the switch was necessary to arm the ascent engine that would lift them off the lunar surface. Worse still, to help ensure that it weighed as little as possible, the Lunar Module contained very few tools. The preparations for the landing had been intense and thorough. But now, if the astronauts were to get home, everyone was going to have to think on their feet.

We will return to Armstrong and Aldrin's predicament in a few moments.

HOW TO EXPECT THE UNEXPECTED

The Apollo team invested vast amounts of time and energy trying to figure out what might go wrong during a mission, and then ensuring that they had a plan for every eventuality. However, it wasn't possible to predict every twist and turn and often the astronauts and mission controllers were forced to deal with the unexpected. Perhaps the most famous of these emergencies occured during a later mission, Apollo 13.

On 11 April 1970, Apollo astronauts Jim Lovell, Fred Haise and Jack Swigert blasted off from the Kennedy Space Center. Two days into the mission, the astronauts heard a loud bang and Swigert uttered his now famous line: 'Houston, we've had a problem here' (misreported in the subsequent Hollywood film about the mission as 'Houston, we have a problem'). The explosion had damaged several pieces of equipment, and within a short period of time the astronaut's Command Module would be running dangerously low on oxygen, water and electricity. The situation looked desperate.

The mission controllers kept their cool, instructing the astronauts to save power by shutting down several electrical systems and severely restricting their water intake. Unfortunately, the resulting impact on the heating reduced the spacecraft's temperature to near freezing and the lack of water subsequently caused astronaut Fred Haise to develop a urinary tract infection. Next, the three astronauts were asked to relocate in their Lunar Module which had its own supply of oxygen and so could be used as a high-tech life raft.

Unfortunately, a new and potentially deadly problem quickly emerged. The Lunar Module's filtration system sucked in air, passed it through a canister of chemicals to remove carbon dioxide and then fed the air back into the spacecraft. The tub of chemicals was designed to cope with two astronauts for around two days, but wouldn't be able to support three astronauts for four days. There was a real risk that the astronauts would die from carbon dioxide poisoning.

The good news was that the Command Module had a similar filtration system and a much larger supply of carbon dioxide-removing chemicals. The bad news was that the canisters used by the Lunar Module's filtration system were cylindrical, while the ones used by the Command Module's system were square. Mission Control was faced with the problem of trying to fit a square canister into a round hole.

The engineers worked hard, eventually coming up with the ultimate space hack. Amazingly, they managed to fit the square canisters into the Lunar Module's filtration system using little more than the cardboard covers from a folder, some plastic sheeting, a roll of duct tape and a towel. The carbon dioxide levels quickly dropped and the astronauts were able to let out a sigh of relief. Eventually, the three brave astronauts made it back home and safely splashed down in the Pacific Ocean. Because the accident had required them to adopt a new flight trajectory, the Apollo 13 astronauts hold the record for being the three human beings to have travelled the furthest from the Earth.

Moonshot Memo
Subject: Invoice for $312,421.24

After the Apollo 13 flight, an engineer who helped build the Lunar Module sent a joke invoice to the company responsible for constructing the Command Module, North American Rockwell.

The invoice was for towing the Command Module back through space (charged at $1 per mile) plus additional costs for charging batteries and carrying an extra passenger – it came to a total of $312,421.24.

North American Rockwell asked its auditor to examine the invoice and then, with their tongue clearly in their cheek, released a statement saying that they had yet to receive payment for the many times that the Command Module had ferried the Lunar Module to the Moon on previous missions.

Apollo 13 wasn't the only time that Mission Control's ingenious improvisational skills saved the day.

On later missions, the astronauts took a Lunar Roving Vehicle with them and used the Moon buggy to explore distant landscapes. During the Apollo 17 mission, astronauts Gene Cernan and Jack Schmitt were on the Moon preparing the Rover and accidentally tore off part of its rear fender (Cernan: 'Oh, you won't believe it. There goes a fender.' Schmitt: 'Oh, shoot!'). The two astronauts used duct tape to reattach the fender and went for a four-hour spin around the lunar surface. Unfortunately, the fix proved less than perfect and problems soon started to emerge. The buggy was kicking up a large amount of Moon dust and, without a proper fender, the dark powder was blackening the astronauts' spacesuits and instrumentation. As a result, both the suits and equipment were absorbing excessive amounts of heat from the Sun and starting to malfunction.

Cernan and Schmitt returned to their Lunar Module and took a scheduled sleep. Mission Control worked on the problem and by the next morning had come up with an improvised hack. They had the astronauts use the duct tape to join four laminated maps together and then clamp the maps to the buggy to act as an impromptu fender. This ingenious act of improvisation proved successful and the rest of the mission ran like clockwork. The astronauts brought their improvised fender back to the Earth, and it's now proudly on display at the Smithsonian's National Air and Space Museum in Washington, D.C.

Flexible thinking brought Apollo 13 safely back to the Earth and provided the Apollo 17 astronauts with a fender for their Moon buggy. But could it help Armstrong and Aldrin lift away from the lunar surface?

THE IMPORTANCE OF FLEX-ABILITY

Apollo 11's Lunar Module used push-button-type circuit breakers. One of the astronauts had knocked off the top of the push-button, leaving the remnants of the switch inside the instrumentation. The button needed to be pushed in to arm the Module's ascent engine. However, as the breaker was part of an electrical circuit, Aldrin was understandably reluctant to place his finger, or anything metal, into the opening.[164]

Then Aldrin had an idea. The Apollo astronauts had been issued with a high-tech space pen that had been specially designed to work in zero gravity. Aldrin didn't much care for it and instead preferred to use felt-tip markers. He suddenly remembered that he had a marker in his shoulder pocket and wondered whether it might do the job. Amazingly, the thin tapered end of the felt tip was a perfect fit and so Aldrin was able to activate the circuit breaker and so prepare the way for the Module's ascent.

The two astronauts then spent the next few hours trying to sleep, with Aldrin curling up on the floor of the Lunar Module and Armstrong lying on the cover of the ascent engine. In another improvisational act, Armstrong had stretched a piece of webbing across the Module to help suspend his legs and so provide a more comfortable sleeping position.

Back on the Earth, some of the mission controllers took the downtime as an opportunity to take a well-earned break. Flight Director Gerry Griffin can remember walking out into the Mission Control car park with a colleague:

It was early morning, but still dark, by the time we completed the Moonwalk. I went outside and it was a beautiful clear July sky in Houston. I looked up, stared at the half Moon, and said to my colleague: 'We've got guys up there, on the Moon, right now. That's pretty neat.' And that was that. We just walked to our cars! I guess there was a feeling of relief and pride. But most of all we wanted to grab some sleep, keep going, and get them back.'[165]

A few hours later the controllers were back at their consoles. Armstrong and Aldrin were woken up. The two astronauts ran through the long list of pre-ascent checks, and Aldrin's pen-based fix seemed to have worked well (Houston: 'For your information, the circuitry looks real fine on that ascent-engine-arm circuit breaker.' Aldrin: 'Roger. I don't think I could get it out now if I wanted to.')[166] There was just one ascent engine and a set amount of fuel. If the propulsion system didn't work, Armstrong and Aldrin would be stranded on the Moon and Collins would be going home alone.

About twenty minutes before lift-off, Aldrin contacted Mission Control and announced: 'Houston, we're No. 1 on the runway.' Michael Collins, the third Apollo 11 astronaut, was alone in the Command Module and reflected on the possibility of his two friends failing to make it back from the Moon. Later describing the experience in his book, *Carrying The Fire*, Collins noted:[167]

'My secret terror for the last six months has been leaving them on the Moon and returning to Earth alone. Now I am within minutes of finding out the truth of the matter. If they fail to rise from the surface, or crash back into it, I am not going to commit suicide; I am coming home.'

Armstrong and Aldrin fired up their ascent engine and the top section of the Lunar Module lifted away from the Moon. As they moved away from the lunar surface, Aldrin had just enough time to look out of the window and see the American flag being blown over by the downdraught from the craft.

Armstrong and Aldrin successfully rendezvoused with Collins in the orbiting Command Module. The astronauts jettisoned the top section of the Lunar Module, journeyed behind the Moon and ignited their engine one more time. There was only one chance at getting the burn right: failure would leave them heading out into deep space or returning rapidly to the surface of the Moon. Fortunately, the firing proceeded to plan and the three astronauts moved away from the Moon and began the long journey home.

Apollo 11 returned on 24 July 1969, piercing the Earth's atmosphere at a remarkable 39,000 kilometres per hour. Emerging into a storm, the astronauts deployed the Command Module's parachutes and splashed down in the Pacific Ocean. The weather was bad and the seas higher than anything they had encountered in training. Eventually a rescue helicopter located them and plucked each astronaut to safety. In Mission Control, flags were waved, cigars were lit and backs were slapped. In May 1961 Present Kennedy had appeared before Congress and declared his audacious and ambitious vision of putting a man on the Moon before the decade was out. Amazingly, the Apollo team had achieved Kennedy's goal and accomplished the seemingly impossible.

President Nixon was on board the aircraft carrier that welcomed Armstrong, Aldrin, and Collins back to Earth, and announced: 'This is the greatest week in the history of the world since creation. As a result of what you have done, the world has never been closer together.'[168]

After spending almost three weeks in quarantine, the three heroes paraded through New York to tumultuous applause and a record tonnage of ticker-tape.

To reduce the Lunar Module's weight on take-off, the astronauts had left several objects on the Moon, including their heavy overshoes, backpacks and urine bags. However, Aldrin held on to his felt tip pen and it currently resides in his office alongside the broken switch. Together, the pen and switch are a lasting tribute to the kind of flexible thinking that was central to the success of the Moon landings.

*

Thriving in the Face of Unexpected Twists and Turns

Let's start with a quick questionnaire. Take a look at the following ten statements. Indicate the degree to which each one describes you on a scale between 1 ('Nope, not me at all') to 5 ('Wow, yes that is definitely me').

		Your score
1	I really enjoy meeting new people at work.	
2	When it comes to holidays, I rarely go back to the same places.	
3	At work I find it easy to think of many different ways of solving a problem.	
4	I rarely find myself thinking of a single word to label my friends, such as 'kind' or 'unreliable'.	
5	When defending a decision at work, I rarely find myself saying something like: 'Well, that's how it's always been done'.	
6	My friends have never described me as stubborn.	
7	When my work colleagues present me with new information about a situation, I often change my mind.	

8	At the weekend, I tend to be spontaneous rather than planning out the days in advance.	
9	At work, I tend to get bored by routine and enjoy new and un-familiar situations and challenges.	
10	I would be fine about making plans to spend an evening with friends and then them suddenly coming up with a different idea about what we should do.	

We will return to the questionnaire later.

In the popular American television show *MacGyver*, secret agent Angus MacGyver combines ordinary objects with his encyclopedic knowledge of science to escape life-and-death situations. In one episode, he is faced with sulphuric acid leaking from a large crack in a giant vat. Realizing that the acid reacts with sugar to create a thick sticky residue, our hero pushes a chocolate bar into the crack and saves the day. In another episode, MacGyver is trapped inside a walk-in freezer. He uses a light bulb to melt some ice, and pours the water into the door lock. When the water re-freezes, it expands and forces the lock open. MacGyver's exploits attracted a large and loyal following, eventually resulting in his name becoming a verb in the *Oxford English Dictionary* ('to make or repair an object in an improvised or inventive way, making use of whatever items are at hand').

This type of improvisational thinking can save lives in the real world. In December 2013, James Glanton and Christina McIntee were driving through an isolated region of northern Nevada with their two young children and their niece and nephew.[169] Unfortunately, their jeep hit a patch of ice, swerved off the road and overturned. With the temperature plunging to below freezing and no mobile telephone signal, the couple began to worry about everyone's safety. Glanton and McIntee thought that their relatives would raise the alarm, but, as they only had their coats for warmth, worried that they might suffer from frostbite or exposure before rescue services arrived. Glanton's improvisational thinking saved the day. First, the couple had the children huddle together in the overturned vehicle. Next, Glanton packed some kindling and wood into the centre of the jeep's spare tire and set fire to it. Finally, Glanton placed several small rocks into the fire, and then ferried the rocks back and forth between the fire and the Jeep. The residual heat from the rocks helped keep everyone warm and when the family was rescued two days later the rescuers were amazed to discover that they had all escaped serious injury.

Psychologists have developed several questionnaires to measure people's level of adaptability, and the statements that you rated at the start of this section are based on their work.[170] In fact, this questionnaire measures your level of adaptability both in the workplace and in your personal life. To discover your adaptability at work, add up the ratings that you gave to the ODD numbered statements (1, 3, 5, 7, 9). Similarly, to find your adaptability in your personal life, add up your ratings to the EVEN numbered statements (2, 4, 6, 8, 10). In both instances a score between 5 and 19 suggests that you have a rigid way of seeing yourself and others, and prefer order to uncertainty. In contrast, a score of between 20 and 25 suggests that you embrace change and find it easy to adapt to new situations.

Armed with these sorts of questionnaires, researchers set out to explore the impact of improvisational thinking in the workplace. In one study, British researchers surveyed over 400 people working in the financial services industry and discovered that psychologically flexible employees felt more in control of their work, were psychologically healthier, and far more productive.[171] Additional work has shown that they are also less stressed, are far less likely to be absent from work, and find it far easier to change careers.[172] These findings, combined with the fact that organizations are now changing at an ever-increasing pace, have resulted in employers valuing the ability to cope with change, adopt new ideas and develop novel ways of working. In fact, one recent survey has shown that the majority of recruitment officers now believe that employees' ability to deal with the unexpected is the most desirable attribute for the future of any organization.[173]

Inspired by these findings, psychologist Ben Fletcher from the University of Hertfordshire began to explore how the same concept affects people's personal lives. Fletcher discovered that many everyday problems are the result of people having an inflexible mindset and being creatures of habit. For instance, if they get into the habit of eating too much and exercising too little, they quickly become overweight. Similarly, if they go to the same sorts of places and chat to the same sorts of people, they struggle to form new friendships and relationship. And if they get used to lighting a cigarette every time they feel stressed, they quickly become a lifelong smoker.

Fletcher teamed up with psychologist Karen Pine to find out what would happen if people adopted a more flexible and improvisational approach to life.[174] In one study, Fletcher and Pine recruited a group of volunteers who wanted to lose weight. Half the volunteers were asked to break their everyday habits by, for example, changing the type of television programmes that they watched, travelling to work using a different route, or taking up a new hobby. The volunteers in the other group were free to follow their favourite diet. Even though those in the 'be more flexible' group weren't asked to cut down on their calories or get more exercise, they had ceased to become creatures of habit and so found it much easier to adopt a healthier diet. As a result, they ended up losing more weight. Other research has shown that the same approach can also help people quit smoking, be happier and even increase their chances of finding their dream job.[175]

We all face the unexpected in both our professional and personal lives. Accidents happen. Markets shift. People change. These uncertainties can knock even the best-laid plans off course, and suddenly throw everyone and everything into uncharted water. The same was true when Armstrong and Aldrin came to blast off from the Moon. They didn't have a plan for what they should do if they accidentally broke off the end of the switch that armed their ascent engine. Instead Aldrin had to live in the moment and work with what he had, and his use of his felt tip pen was clever and effective.

Unfortunately, when it comes to improvisational thinking, Aldrin is the exception rather than the rule. Most people tend to be creatures of habit and carry out the same routines day after day. However, the good news is that it is easy to shift into a more flexible way of looking at the world. It's just a question of engaging in a spot of mental yoga, meeting a mysterious stranger, and being willing to risk it all, on the roll of a die.

MENTAL YOGA

Go to yoga classes and you will become more physically flexible. Similarly, doing something different makes you more mentally flexible. In one experiment, for instance, Dutch psychologist Simone Ritter asked volunteers to prepare a sandwich involving butter and chocolate chips.[176] This dish is well known in the Netherlands and would usually involve putting the bread on a plate, buttering it and then sprinkling on the chocolate chips. One group of volunteers was asked to make the sandwich in the usual way, while a second group was asked to mix things up by laying the chocolate chips on a plate, buttering the bread and then pushing the bread (butter-side-down) onto the chocolate chips. Compared to those who went about their sandwich-making in the usual way, the strange sandwich makers scored much higher on tests of flexible thinking.

The same idea applies to everyday life. Psychologist Professor Ben Fletcher has devoted much of his academic career to helping people to become more adaptable. Much of his work has involved repeatedly taking people out of their comfort zones and having them encounter the unexpected. Fletcher has developed a training programme called 'Do Something Different', in which volunteers are asked regularly to expose themselves to novel experiences. He has worked with many organizations and individuals, his work illustrating that regularly encountering the unexpected helps develop a more flexible way of thinking.

Make the most of this approach by spending one day a week

doing something you have never done before. Here are a few ideas that will help make the exercise especially effective.

Habit busting: Try changing one of your most ingrained habits. For instance, you might stop watching television for a week, try a new type of food, listen to a novel type of music, go to a museum or art gallery that you have never visited before, develop a different workout at the gym or buy something in a shop that you have never been to before. If you enjoy being in control, allow your partner or friend to plan your day without consulting you or if you have a habit of saying 'no' to opportunities, spend a week saying 'yes' to everything that comes your way.

Get in touch with your inner clown: You are especially adaptable when you are relaxed and having fun. To get in touch with your inner adaptable clown, try composing a poem, writing a short story in fifteen minutes, scribbling on a piece of paper and then turning the scribble into a drawing, finishing a crossword without reading the clues, moonwalking across the room, having a long conversation without using the word 'I', watching a film that you think you will hate or building a sofa-fort with your kids and spending the night in it.

It takes all sorts: Research shows that creative entrepreneurs tend to network with a diverse group of people, and that the diversity of perspectives and knowledge helps them to see the world in a more flexible way.[177] Similarly, in your own life, try to mix things up. Go out of your way to create a diverse range of acquaintances, colleagues and friends. The same goes for teamwork. A few years ago, researchers examined the relationship between teamwork and people's approach to developing new products and services.[178] Some people have a tendency to create completely new ideas ('innovators'), while others are more comfortable adapting existing ones ('adapters'). The researchers measured the effectiveness of teams of either purely innovators, purely adapters or a mixture of the two styles of thinking. The teams that contained a mixture of innovators and adapters were significantly more effective than those consisting of just one style.

Travel

Finally, remember that travel broadens the mind.[179] A few years ago, Professor William Maddux, from the Kellogg School of Management at Northwestern University, asked a group of volunteers how long they had spent living abroad and also tested their ability to think of different ways of using everyday objects. Diversity paid dividends, with the results revealing that the longer the volunteers had spent abroad, the more flexible their minds were. And you don't have to travel far to get the advantage. In another study, Maddux asked French students to merely think about a time when they had lived abroad and even this put them into a more flexible state of mind.

Practise mental yoga on a daily basis and you will become used to dealing with the unexpected, and so adapt and thrive in the face of change.

MEETING A MYSTERIOUS STRANGER

To help develop a more adaptive and improvisational attitude, try this fun challenge.

Imagine that you are having a quiet drink in a bar. A mysterious stranger walks up to your table, reaches into their pocket and places three objects on the table: a candle, a book of matches and a small cardboard box containing some drawing pins. The stranger lights the candle, and challenges you to safely fix the lit candle to the wall, but in such a way that the hot wax won't drip onto the floor.

How would you try to win the challenge?

_____

_____

Maybe you thought of using the drawing pins to fix the candle to the wall or came up with using some of the hot wax from the candle to hold it in place. Unfortunately, neither method would work because the candle is far too close to the wall and the hot wax will end up all over the floor. Perhaps you thought about contacting your engineering friend and having them develop a new form of candle-holding device or about going behind the bar and seeing if you can find an implement that might prove helpful.

If you did come up with any of these ideas, read the challenge again and think about other ways of solving the problem. The answer is surprisingly straightforward, but does require the type of thinking that got MacGyver out of many a sticky situation.

This challenge, known as 'Duncker's Candle Problem', was

created by psychologist Karl Duncker in the 1940s and has been presented to thousands of people in hundreds of experiments. Have you come up with a solution yet? The most elegant approach involves emptying the box of drawing pins, using a couple of them to fix the cardboard box to the wall and then standing the candle on the box.

Only around 40% of people manage to come up with this simple solution. Why? Because ever since they were a child they have seen boxes holding objects and, without them knowing it, their brains have become fixed and unable to see what's right in front of their eyes.

Thinking in an adaptive and flexible way is often about working with what you already have to hand. To use a well-known analogy, when you are hungry, it's less a case of heading out to the nearest restaurant and more about seeing what you can create from the ingredients that you already have in your cupboard.

Several organizations have used this approach to create new forms of technology. Shunning expensive, high-tech solutions, these so-called 'frugal innovators' often focus on adaptive ways of using resources that are already to hand. Take, for example, the soda bottle light bulb. In many tropical countries, the poor often have no choice but to live in small dark rooms that shield them from the torrential rain and the blazing sun. As a result, they are forced to rely on expensive electric lights for illumination during the day. However, in the Philippines, aid workers have developed a novel way of bringing light into these spaces. They fill a clear one-litre plastic bottle with water and add a small amount of bleach to prevent mould. Next, they cut a hole in the roof and wedge the bottle into the hole, ensuring that half the bottle sticks above the roof. The light from the tropical sun then hits the top half of the bottle and the water refracts the light into the room. Using just a discarded soda bottle, some water and a spot of bleach, the aid workers have created the safest, cheapest, 55-watt light bulb in the world.

In another example, Stanford Professor of Bioengineering Manu Prakash has created the world's most economical microscope. His origami-based Foldscope can be assembled from a pre-printed sheet of card in just a few minutes and uses an inexpensive glass lens. The resulting device weighs just eight grams and provides enough magnification to spot various dangerous bacteria and parasites. Now people in the developing world don't need to travel long distances or wait months to see a specialist physician. Instead, for just a few pence they can be tested and diagnosed for treatment.

The next time you are tempted to invest time or money in a new venture, take a look at whether you can be far more frugal and flexible by adapting what you already have to hand.

RISKING IT ALL ON THE ROLL OF A DIE

Published in 1971, Luke Rhinehart's best-selling novel *The Dice Man* has become a modern cult classic. The book tells of a psychiatrist who starts to make major life decisions on the basis of the roll of dice. The book has its roots in reality. As a boy, Rhinehart was shy and started to use the dice technique to make himself do things that he was otherwise too nervous to do. Happy with the results, dicing became part of his life and he used the technique to help him overcome procrastination and push him out of his comfort zones.

The role of the dice influenced his life in unexpected ways. One day, for instance, Rhinehart was driving home when he saw two nurses walking along. He decided that if his die came up with an odd number he would offer the nurses a lift. Rhinehart rolled his die and soon found himself chatting to the nurses in his car. The journey went well, with Rhinehart becoming smitten with one of the women and eventually marrying her.

Over the years, people across the world have used the dicing technique to add a sprinkling of unpredictability to their lives. Entrepreneur Richard Branson admitted using the technique when he was a young man. The dice initially dictated that he should make a loud scream every hour, on the hour.[180] A few rolls later, Branson discovered that he had to fly to Finland to see a band called Wigwam. Later that day, Branson was in Helsinki watching Wigwam perform an acoustic number. As the clock struck 10 p.m., Branson stuck to his guns and let out a huge scream. The audience was confused and the band was less than delighted. Unrepentant, Branson reprised his screaming during the encore. After twenty-four hours, Branson decided that it was all getting out of hand and stopped his experiment. Nevertheless, he says he has been influenced by the dice many times in his career.

Add a random element to life by dicing. When you have a decision to make, write down six possible scenarios. Maybe you are trying to decide what to do with the day, where to go one evening, or what television programme to watch. If you feel brave, maybe you will be prepared to roll the die and decide who you should spend more time with, what you should give up for a year, or which new language you will learn. Only choose options that you are really prepared to carry out. Write down your list of possible scenarios and then number them one to six and get ready to roll a die. Just before you do, make a promise. Promise yourself that you will carry out whichever scenario is selected. There's no exchanging the scenario for another one or backing out.

Now roll the die, enjoy the experience and find out what it's like to be on the other side of rigid. You may never find yourself having to use a felt tip pen to blast away from the Moon, but your new-found adaptable and improvising mind will prove an equally vital asset in everyday life.

SUMMARY

Being able to cope with the unexpected, improvise and adapt is vital to success. To become a more flexible thinker:

– Try a spot of mental yoga. Do something different. Maybe take up a new hobby or interest, meet new people, and change your route to work.

– Before investing time or money in a new venture, be far more frugal and flexible by seeing if you can use resources that you already have to hand.

– Risk it all on the roll of a die. Jot down six actions or potential solutions to a problem, roll a die and embrace the power and fun of the unexpected.

MISSION ACCOMPLISHED

America's space programme had achieved the seemingly impossible. It was just seven years since Kennedy's historic speech in Houston, yet humanity had travelled to the Moon and returned safely back to Earth. The mission controllers sat at the heart of the entire enterprise. People like Jerry Bostick, who had grown up in rural Mississippi and helped to plot the trajectories that took the Apollo spacecraft to the Moon; twenty-six-year-old Steve Bales, who had begun his career giving tours around the Manned Spacecraft Center, and ended up monitoring the Lunar Module during Armstrong's historic decent; Ed Fendell, who initially studied for a degree in merchandizing and went on to supervise the systems that allowed the astronauts to communicate with their fellow Earthlings, and Chris Kraft, who grew up in a tough town and laid the foundations for Mission Control and quietly led the team to success.

During our time together, we have explored the eight psychological principles that I believe underpin Mission Control's astonishing achievement. We have discovered the motivational power of passion and the importance of innovation. We have seen how self-belief is crucial when it comes to making a start, and how learning to fail is vital to victory. We have found out how conscientiousness forms the bedrock for success, and how courage provides a springboard for progress. And finally, we have explored the positive form of pessimism that underpins preparedness, and discovered the importance of being flexible in the face of unexpected twists and turns.

It's time for one final exercise. In the same way that the mission controllers often produced crib sheets to remind them of important information, here's a breakdown of the eight principles and some of the most important techniques, at play. As you go through them, think about the degree to which they apply to you (or, as the controllers might say, whether they are 'Go' or 'No Go') and so which ones require your attention in the future. . . .

PASSION

Just as Kennedy energized a nation with his audacious vision of going to the Moon, be passionate about your aims and ambitions.

- Create a grand goal, set a dramatic deadline, or find a way of being the first to do something.

- Inject a sense of purpose into any activity by asking that one simple question: 'How does this help others?', or generate your own space race by creating a sense of competition.

INNOVATION

In the same way that John Houbolt created an innovative mission plan, come up with lots of original ideas and ensure that the best one wins.

- Harness the power of 'vice versa' thinking by doing the opposite to everyone else (remember how most engineers favoured one giant rocket and Houbolt opted for several smaller spacecraft).

- Use the 'less is more' rule by imagining what you would do if you only had half the resources, time, or funds.

SELF-BELIEF

The mission controllers were so young that they didn't know that it was almost impossible to get to the Moon in just a few years. Develop the same level of self-belief.

– Use the power of small wins by breaking your goal into smaller stages, and celebrating after you achieve each step.

– If you experience self-doubt, spend a few moments thinking about your most impressive accomplishments to date.

LEARN HOW TO FAIL

The tragic Apollo 1 fire resulted in the Apollo team being more open about mistakes and learning from errors. Similarly, accept tricky challenges, admit your mistakes, and see failure as an opportunity for growth.

– Follow in Dale Carnegie's footsteps, and keep a list of all the damned foolish things that you have done and what you learned from them.

– Remember to develop a growth mindset by using the magic word 'yet' ('I am not great at going to the gym, yet').

RESPONSIBILITY

Adopt Apollo's 'It won't fail because of me' attitude by taking responsibility for what you do and what you don't do.

– Overcome procrastination. Remember the mantra: 'Don't do nothing, just because you don't have time to do everything you want to do.' Create precise deadlines ('I will email you by 3 p.m. tomorrow').

– Avoid over-commitment. When you are asked if you want to do something at a future date, say to yourself: 'Would I want to do it tomorrow?' If the answer is 'no', decline the request.

COURAGE

Flight Director Glynn Lunney once memorably remarked: 'If you're going to go to the Moon, sooner or later you've got to go to the Moon.' Find the courage to stop talking and start acting.

– Assess the risks, and remember Kennedy's words of wisdom: 'There are risks and costs to action. But they are far less than the long range risks of comfortable inaction.'

– Take risks, but don't be reckless and be wary of pushing on with an under-taking just because you have invested time, energy or money (remember, sometimes we don't need to go to the Moon today).

PREPAREDNESS

In the same way that Mission Control rehearsed for every eventuality, be fully prepared.

– Use 'what if . . .' thinking to develop contingency plans for likely scenarios.

– Carry out a 'pre-mortem' by imagining that your project has already failed, determining what went wrong and figuring out how you can prevent major problems.

FLEXIBILITY

Buzz Aldrin's flexible use of a felt-tip pen helped ensure that the Lunar Module blasted away from the surface of the Moon. Similarly, when the unexpected strikes, be ready to improvise and adapt.

– Regularly do something different. Try a new type of food, take up a new hobby or interest, meet new people, or change your route to work.

– Be prepared to risk it all on the roll of a die. Jot down six actions or potential solutions to a problem, roll a die and go with the flow.

My interviews with the controllers provided one final surprise. Being successful is one thing, but being able to handle the consequences of that success is quite another. Over the years I have spent a considerable amount of time with many highly successful CEOs, leaders, and celebrities. Triumph can often go to people's heads, causing them to become boastful and self-important. Once again, the Apollo programme provides a guiding light. Despite playing a central role in humanity's greatest achievement, the mission controllers are by far the most humble group of people that I have ever had the pleasure of interviewing. Time and again, they would choose the word 'we' over 'I', and were always quick to point out the important role played by their colleagues. Many of them spoke about the honour of being given the opportunity to work on the space programme and the good luck to have lived at a time when the entire nation supported such a big and bold endeavour.

This type of humility is often misperceived as a form of low self-esteem. In fact, research shows that the opposite is true. Humble people tend to feel especially secure, and are therefore happy to play down their own achievements and share the limelight. For them, success is not about boosting their status or puffing their ego, but rather about putting their accomplishments in perspective, and recognizing the role played by factors outside themselves, such as their upbringing, luck, and other people.

Recent research has shown that humble people tend to be more altruistic, forgiving, likeable, grateful, and cooperative.[181] Perhaps not surprisingly, this results in them having more friends than most, and being especially likely to be in loving and secure relationships. The same applies in the workplace. Humility is vital when it comes to teamwork, and so positively related to job satisfaction and productivity. Similarly, humility underpins excellence in leadership, and is one of the most important factors in moving organizations from good to great.

The mission controllers' sense of humility may have been a result of their modest backgrounds, the societal norms in the 1950s and 1960s or that they had to work as a team in order to get to the Moon. Whatever the cause, their humility is both admirable and heart-warming. Unfortunately, such modesty seems to be increasingly rare in modern-day society, with several researchers documenting a dramatic increase in self-aggrandisement and narcissism. Perhaps fuelled by social media, the world now appears to be full of people who are eager to put themselves centre stage.

Fortunately, it doesn't take much to change. When success comes your way, spend a few moments thinking about how the mission controllers speak about their astonishing achievement. Choose 'we' over 'I'. Reflect on how your success was due to the support of your friends, partner, teachers, parents, family, and colleagues. Acknowledge the role of good luck, your upbringing, and the situation in which you found yourself.

A vast number of technological benefits flowed from the Apollo programme. From home insulation to shock-absorbing trainers, memory-foam mattresses to scratch-resistant lenses, and flame-resistant clothing to integrated circuits, your everyday life is a direct result of the Moon missions. However, one of the most enduring benefits has been psychological rather than technological.

On Christmas Eve 1968, the Apollo 8 astronauts were in orbit around the Moon. At one point they looked out from the capsule window, saw the Earth rising above the Moon and quickly snapped several colour photographs. The best of the images – known as Earthrise – was taken by William Anders and has become one of the most famous and frequently reproduced photographs in history. When people look at the Earthrise photograph they often experience a sudden shift in perspective. Some look at the beautiful blue marble suspended in space, compare it to the lifeless and barren lunar landscape and appreciate how fortunate they are to be living on the Earth. For others, the photograph shows that humanity is totally alone and adrift in a vast universe; it provides a vivid illustration of the banality of our terrible wars and cruelties. For many, the photograph reveals just how small and fragile the Earth really is, prompting an increased sense of planetary stewardship. Indeed, the Earthrise photograph, along with similar images from the Apollo missions, are widely credited with driving forward the environmental movement. As astronaut Bill Anders once memorably phrased it: 'We came all this way to explore the moon, and the most important thing is that we discovered the Earth.'

In the same way that photographs of the Earth from space caused millions of people to alter how they saw themselves and the world, so the mindset that put us on the Moon allows us to view success in a radically new way. When you think about high achievers, you might be tempted to conjure up images of Olympians who, through their genetic make-up and extremely strict training regime, have earned a place on the podium. Or hard-headed, tough-talking CEOs, whose only concern is to boost the bottom line, multimillionaires, from privileged backgrounds, who have benefited from their social connections and inherited wealth or entrepreneurs who have made their fortune by taking huge risks, and building one empire after another.

The mission controllers don't fit into any of these moulds and so rewrite the story of success. They were a group of ordinary people, from modest backgrounds, who achieved a seemingly impossible goal for the good of humanity. They are testament to a new form of success. Oh, and throughout it all, they managed to stay humble.

Each time that you look up at the Moon, remember their inspirational story.

Against all of the odds, they got there.

You can, too.

NOTES

1   This is the original text of Kennedy's speech at Rice University. When he delivered it, he moved away from the written text, stating: *'We choose to go to the moon in this decade and do the other things, not because they are easy, but because they are hard, because that goal will serve to organize and measure the best of our energies and skills, because that challenge is one that we are willing to accept, one we are unwilling to postpone, and one which we intend to win, and the others, too.'* Original text from the John F. Kennedy Presidential Library and Museum.

2   There has been a small amount of work in this area. For instance, Flight Director Gene Kranz produced a list of six items known as the 'Foundations of Mission Control' (Discipline, Competence, Confidence, Responsibility, Toughness and Teamwork); historian Andrew Chaikin delivered a talk on Apollo project management at the Goddard Space Flight Center in July 2012, and in 1989 several senior officials involved in the Apollo programme came together to discuss management and leadership (Logsdon, J. M. (ed.) (1989). *Managing the Moon Program: Lessons Learned from Project Apollo: Proceedings of an Oral History Workshop.* Monographs in Aerospace History. Number 14. NASA.

3 For additional information about Sputnik, see: Boyle, R. (2008), 'A Red Moon over the Mall: The Sputnik Panic and Domestic America'. *Journal of American Culture*, 31: 373–82.

4 Reproduced with kind permission from the Bentley Historical Library, University of Michigan.

5 Boyle, R. 'A Red Moon over the Mall', 374.

6 Ibid 375.

7 Murray, C. A. & Cox, C. B. (1989). *Apollo: The Race to the Moon.* Simon & Schuster: New York, 23–4.

8 Boyle, R. 'A Red Moon over the Mall', 378.

9 Logsdon, J. M. (2010). *John F. Kennedy and the Race to the Moon.* Palgrave Studies in the History of Science and Technology: London.

10 Interview with author.

11 Interview with author.

12 Cited in Logsdon, J. M. (ed.) (1989). *Managing the Moon Program.*

13 Cited in Swanson, G. (ed.) (2012). *Before This Decade Is Out.* Dover Publications: New York.

14 Interview with author.

15 Interview with author.

16 For an overview of this work, see: Vallerand, R. J. (2015). *The Psychology of Passion: A Dualistic Model.* Open University Press: New York.

17 Schellenberg, B., & Bailis, D. (2015). 'Can Passion Be Polyamorous? The Impact of Having Multiple Passions on Subjective Well-Being and Momentary Emotions'. *Journal of Happiness Studies*, 16 (6), 1365–81.

18 Howatt, W. A. (1999). 'Journaling to Self-evaluation: A Tool for Adult Learners'. *International Journal of Reality Therapy*, 182, 32–4.

19 Campbell, E. T. (1970) '"Give Ye Them to Eat": Luke 9:10–17'. In 'Ser-

mons from Riverside'. The Publications Office, The Riverside Church: New York.

20 Bunderson, J., & Thompson, J. (2009). 'The Call of the Wild: Zoo-keepers, Callings, and the Double-edged Sword of Deeply Meaningful Work'. *Administrative Science Quarterly*, 54, 32–57.

21 Grant, A. M., Campbell, E. M., Chen, G., Cottone, K., Lapedis, D., & Lee, K. (2007). 'Impact and the Art of Motivation Maintenance: The Effects of Contact with Beneficiaries on Persistence Behavior'. *Organizational Behavior and Human Decision Processes*, 103, 53–67.

22 For a review of this work, see: Wrzesniewski, A., LoBuglio, N., Dutton, J. E., & Berg, J. M. (2013). 'Job Crafting and Cultivating Positive Meaning and Identity in Work', in Arnold B. Bakker (ed.) *Advances in Positive Organizational Psychology* (*Advances in Positive Organizational Psychology*, Volume 1). Emerald Group Publishing Limited, 281–302.

23 Triplett, N. (1898). 'The Dynamogenic Factors in Pacemaking and Competition'. *American Journal of Psychology*, 9, 507–33.

24 Murayama, K., and Elliot, A. J. (2012). 'The Competition–Performance Relation: A Meta-analytic Review and Test of the Opposing Processes Model of Competition and Performance'. *Psychological Bulletin*, 138, 1035–70.

25 Kilduff, G. J. (2014). 'Driven to Win: Rivalry, Motivation, and Performance'. *Social Psychological and Personality Science*, 5, 944–52.

26 For biographical information about Wernher von Braun, see: Ward, R. (2009). *Dr. Space: The Life of Wernher Von Braun*. Naval Institute Press: MD.

27 For additional information about von Braun and the Nazis, see: Neufeld, M. J. (2013). *The Rocket and the Reich: Peenemunde*

and the Coming of the Ballistic Missile Era. Smithsonian Books: Washington DC.

28   For additional information about von Braun's public-facing work, see: Scott, D. M. & Jurek, R. (2014). *Marketing the Moon*. MIT Press: Cambridge, MA.

29   For additional information about von Braun's decisions about how best to go to the Moon, see: Neufeld, M. J. (2008). 'Von Braun and the Lunar-Orbit Rendezvous Decision: Finding a Way to Go to the Moon'. Acta Astronautica 63. 540–50.

30   For additional information about Houbolt's work, see: Hansen, J. R. (1995). *Enchanted Rendezvous: John C. Houbolt and the Genesis of the Lunar-Orbit Rendezvous Concept. Monographs in Aerospace History*, No. 4. NASA History Division: Washington DC.

31   Wimbiscus, B. (2014). 'John C. Houbolt, a Joliet Native, Remembered for Helping Put Man on the Moon'. *Herald-News*, 22 April.

32   Yardley, W. (2014). 'John Houbolt, NASA Innovator Behind Lunar Module, Dies at 95'. *New York Times*, 27 April.

33   Hansen, J. R. (1995). *Enchanted Rendezvous*.

34   See, for example: Luchins, A. S. (1942). *Mechanization in Problem Solving: The Effect of Einstellung. Psychological Monographs*. 54 (6): i–95. Rokeach, M. (1948). 'Generalized Mental Rigidity as a Factor in Ethnocentrism'. *Journal of Abnormal and Social Psychology*, 43(3), 259–78.

35   Gregg, A. P., Mahadevan, N., & Sedikides, C. (2016). 'The SPOT effect: People Spontaneously Prefer Their Own Theories'. *Quarterly Journal of Experimental Psychology*, Section B, 70, 996–1010.

36   For a review of this work, see: Kyung, H. K. (2011), 'The Creativity Crisis: The Decrease in Creative Thinking Scores on the Torrance Tests of Creative Thinking', *Creativity Research Journal*, 23(4), 285–95.

37   Cooper, B. L., Clasen, P., Silva-Jalonen, D. E., & Butler, M. C. (1999). 'Creative Performance on an In-basket Exercise: Effects of Inoculation against Extrinsic Reward'. *Journal of Managerial Psychology*, 14(1), 39–57.

38   Scopelliti, I., Cillo, P., Busacca, B., & Mazursky, D. (2014). 'How Do Financial Constraints Affect Creativity?' *Journal of Product Innovation Management*, 31(5), 880–93.

39   Sio, U. N., & Ormerod, T. C. (2009). 'Does Incubation Enhance Problem Solving? A Meta-analytic Review'. *Psychological Bulletin*, 135, 94–120.

40   Gilhooly, K. J., Georgiou, G. J., Garrison, J., Reston, J. D., & Sirota, M. (2012). 'Don't Wait to Incubate: Immediate Versus Delayed Incubation in Divergent Thinking'. *Memory and Cognition*, 40, 966–75.

41   Oppezzo, M., & Schwartz, D. L. (2014). 'Give Your Ideas Some Legs: The Positive Effect of Walking on Creative Thinking'. *Journal of Experimental Psychology: Learning, Memory, and Cognition*, 40(4), 1142–52.

42   Wagner, U., Gais, S., Haider, H., et al. (2004). 'Sleep Inspires Insight'. *Nature* 427: 352–5.

43   Mednick, S. C., Kanady, J., Cai, D., & Drummond, S. P. A. (2008). 'Comparing the Benefits of Caffeine, Naps and Placebo on Verbal, Motor, and Perceptual Memory'. *Behavioral Brain Research*, 193, 79–86.

44   Huang, Y., Choe, Y., Lee, S., Wang, E., Wu, Y., & Wang, L. (2018). 'Drinking Tea Improves the Performance of Divergent Creativity'. *Food Quality and Preference* 66, 29–35.

45   https://www.airspacemag.com/space/first-up-1474936/

46   For additional information, see: Wolfe, T. (1979). *The Right Stuff*. Farrar, Straus and Giroux: New York. Conrad, N., & and Klausner, H. A. (2005). *Rocketman: Astronaut Pete Conrad's Incredible Ride to the Moon and Beyond*. Penguin Books: London.

47 Glenn, J., & Taylor, N. (1985). *John Glenn: A Memoir*. Bantam Books: New York.

48 Kranz, G. (2009). *Failure Is Not an Option: Mission Control from Mercury to Apollo 13 and Beyond*. Simon & Schuster: New York.

49 https://history.nasa.gov/SP-4201/ch11-4.htm

50 Kraft, C. (2001). *Flight: My Life in Mission Control*. Dutton Books: New York.

51 Burgess, C. (2015). *Friendship 7: The Epic Orbital Flight of John H. Glenn, Jr*. Springer International Publishing: New York.

52 Bostick, J. (2016). *The Kid from Golden. From the Cotton Fields of Mississippi to NASA Mission Control and Beyond*. iUniverse: Bloomington, IN.

53 Interview with author.

54 Interview with author.

55 For an overview of this theory, see: Bandura, A. (1977). 'Self-efficacy: Toward a Unifying Theory of Behavioral Change'. *Psychological Review*, 84, 191–215.

56 Bandura, A. (1997). *Self-efficacy: The Exercise of Control*. Freeman: New York.

57 Amabile, T. M., & Kramer, S. J. (2011). *The Progress Principle: Using Small Wins to Ignite Joy, Engagement, and Creativity at Work*. Harvard Business Review Press: Cambridge, MA.

58 Interview with author.

59 Sparrow, K. R. (1998). 'Resiliency and Vulnerability in Girls During Cognitively Demanding Challenging Tasks'. PhD Thesis, Florida State University: Tallahassee, FL.

60 Although there is some debate about who originally came up with the plot of 'The Little Engine That Could', the most popular version of the story was published in 1930 by Platt & Munk, and this retelling was credited to Watty Piper (pseudonym for Arnold Munk).

61 Fowler, J. H., & Christakis, N. A. (2008). 'Dynamic Spread of Happiness in a Large Social Network: Longitudinal Analysis over 20 Years in the Framingham Heart Study'. *BMJ*, *337*, a2338.

62 Damisch, L., Stoberock, B., & Mussweiler, T. (2010). 'Keep Your Fingers Crossed! How Superstition Improves Performance'. *Psychological Science*, 21(7), 1014–20.

63 Keller, H. (1903). *Optimism*. Crowell & Company: New York.

64 Bannister, R. (2014). *Twin Tracks: The Autobiography*. The Robson Press: London.

65 Schrift, R. Y., & Parker, J. R. (2014). 'Staying the Course: The Option of Doing Nothing and Its Impact on Postchoice Persistence'. *Psychological Science*, 25(3), 772–80.

66 Boomhower, R. E. (2004). *Gus Grissom: The Lost Astronaut* (Indiana Biography Series). Indiana Historical Society: Indianapolis.

67 Grissom, B., & Still, H. (1974). *Starfall*. Thomas Crowell Company: New York.

68 White, M. C. (2006). 'Detailed Biographies of Apollo I Crew – Gus Grissom. NASA History'. https://history.nasa.gov/Apollo204/zorn/grissom.htm

69 Howell, E. (2018). 'How John Young Smuggled a Corned-Beef Sandwich into Space'. 10 January. https://www.space.com/39341-john-young-smuggled-corned-beef-space.html

70 White, M. C. (2006). 'Detailed Biographies of Apollo I Crew – Gus Grissom'. NASA History'.https://history.nasa.gov/Apollo204/zorn/grissom.htm

71 Interview with author.

72 Interview with author.

73 Grissom, V. I. (1968). *Gemini: A Personal Account of Man's Venture Into Space*. Macmillan: New York.

74 United States. (1967). Report of Apollo 204 Review Board to the Administrator, National Aeronautics and Space Administration. *Washington, DC: National Aeronautics and Space Administration.*

75 Kranz, G. (2009). *Failure Is Not an Option.*

76 Houston, R., & Heflin, M. (2015). *Go, Flight! The Unsung Heroes of Mission Control.* University of Nebraska Press: Lincoln, NB, 33.

77 Interview with author.

78 Logsdon, J. M. (ed.) (1999). 'Managing the Moon Program: Lessons Learned From Project Apollo: Proceedings of an Oral History Workshop'. Monopolies in Aerospace History No. 14. NASA.

79 Dweck, C. S. (2012). *Mindset: How You Can Fulfil Your Potential.* Constable & Robinson Limited, London. There have recently been several meta-analyses of this work, some of which have suggested that the effects are relatively small. See, for example: Sisk, V. F., Burgoyne, A. P., Sun, J., Butler, J. L., & Macnamara, B. N. (2018). 'To What Extent and Under Which Circumstances Are Growth Mind-Sets Important to Academic Achievement? Two Meta-Analyses'. *Psychological Science*, 29 (4), 549–71.

80 Blackwell, L., Trzesniewski, K., & Dweck, C. S. (2007). 'Implicit Theories of Intelligence Predict Achievement Across an Adolescent Transition: A Longitudinal Study and an Intervention'. *Child Development*, 78, 246–63.

81 Ehrlinger, J., Burnette, J. L., Park, J., Harrold, M. L., & Orvidas, K. (In press). 'Incremental Theories of Weight Predict Lower Consumption of High-Calorie, High-Fat Foods'. *Journal of Applied Social Psychology.*

82 Keating, L. A., & Heslin, P. A. (2015). 'The Potential Role of Mindsets in Unleashing Employee Engagement'. *Human Resource Management Review*, 25, 329–41. Heslin, P. A., & Keating, L. A. (2017). 'In

Learning Mode? The Role of Mindsets in Derailing and Enabling Experiential Leadership Development'. *The Leadership Quarterly*, 28, 367–84. Heslin, P. A., Latham, G. P., VandeWalle, D. (2005). 'The Effect of Implicit Person Theory on Performance Appraisals'. *Journal of Applied Psychology*, 90, 842–56. *Why Fostering a Growth Mindset in Organizations Matters*. Report published by Senn Delaney. http://knowledge.senndelaney.com/docs/thought_papers/pdf/stanford_agilitystudy_hart.pdf

83   For a review of the history of this type of puzzle, see: Kullman, D. (1979). 'The Utilities Problem'. *Mathematics Magazine*, 52(5), 299–302.

84   Houston, R., & Heflin, M. *Go, Flight!*, 21.

85   Watts, S. (2013). *Self-Help Messiah*. Other Books: New York.

86   This type of exercise has been developed and used by several researchers exploring the concept of growth and fixed mindsets. See, for instance: Aronson, J., Fried, C., & Good, C. (2002). 'Reducing the Effects of Stereotype Threat on African American College Students by Shaping Theories of Intelligence'. *Journal of Experimental Social Psychology*. 38, 113–25.

87   Mueller, C. M., & Dweck, C. S. (1998). 'Intelligence Praise Can Undermine Motivation and Performance'. *Journal of Personality and Social Psychology*, 75, 33–52.

88   For additional information about this idea, see Carol Dweck's TED talk, 'The power of believing that you can improve'.

89   Burgess, C. (2014). *Liberty Bell 7: The Suborbital Mercury Flight of Virgil I. Grissom*. Springer International Publishing Switzerland. Chapter 2, 'An astronaut named Gus'.

90   Personal communication.

91   http://www.philstar.com/business-life/2013/0(6)/17/954(6)98/out-world-teamwork-lessons-nasa

92   Farmer, G., & Hamblin, D. J. (1970). *First on the Moon: A Voyage with Neil Armstrong, Michael Collins and Edwin E. Aldrin, Jr.* Little Brown: Boston, MA., 77.

93   Interview with author.

94   Interview with author.

95   Interview with author.

96   Interview with author.

97   Interview with author.

98   Farmer, G. & Hamblin, D. J. *First on the Moon*, 76.

99   Ibid.

100 Schirra, W., & Billings, R. N. (1988). *Schirra's Space*. Quinlan Books: Boston, MA.

101 https://www.wallyschirra.com/gemini.htm

102 Decker, W. H., & Rotondo, D. M. (1999). 'Use of Humor at Work: Predictors and Implications'. *Psychological Reports*, 84 (3), 961–8.

103 Minton, H. L. (1988). *Lewis M. Terman: Pioneer in Psychological Testing. American Social Experience Series*. New York University Press: New York.

104 Borghans, L., Golsteyn, B. H. H., Heckman, J. J., & Humphries, J. E. (2016). 'What Grades and Achievement Tests Measure'. *Proceedings of the National Academy of Sciences*, 113 (47) 13354–9. Duckworth, A. L., Weir, D., Tsukayama, E., & Kwok, D. (2012). 'Who Does Well in Life? Conscientious Adults Excel in Both Objective and Subjective Success'. *Frontiers in Psychology*, 3, 356, 1–8. Poropat, A. E. (2014). 'Other-rated Personality and Academic Performance: Evidence and Implications'. *Learning and Individual Differences*, 34, 24–32.

105 Judge, T. A., Higgins, C.A., Thoresen, C. J., & Barrick, M. R. (1999). 'The Big Five Personality Traits, General Mental Ability, and Career Success Across The Life Span'. *Personnel Psychology*, 52, 621–52. Rob-

erts, B. W., Walton, K., & Bogg, T. (2005). 'Conscientiousness and Health across the Life Course'. *Review of General Psychology*, 9, 156–68. Roberts, B. W., & Bogg, T. (2004). 'A 30-Year Longitudinal Study of the Relationships between Conscientiousness-Related Traits, and the Family Structure and Health-Behavior Factors that Affect Health'. *Journal of Personality*, 72, 325–54.

106 Solomon, B. C., & Jackson, J. J. (2014). 'The Long Reach of One's Spouse: Spousal Personality Influences Occupational Success'. *Psychological Science*, 25, 2189–98.

107 Jackson, J. J., Wood, D., Bogg, T., Walton, K. E., Harms, P. D., & Roberts, B. W. (2010). 'What Do Conscientious People Do? Development and Validation of the Behavioral Indicators Of Conscientiousness (BIC)'. *Journal of Research in Personality*, 44, 501–11.

108 Rotter, J. B. (1966) 'Generalized Expectancies for Internal Versus External Control of Reinforcement'. *Psychological Monographs*, 80, 1–28.

109 Ng, T. W. H., Sorensen, K. L. & Eby, L. T. (2006), 'Locus of Control at Work: A Meta-Analysis'. *Journal of Organizational Behavior*, 27, 1057–87.

110 Mendoza, J. C. (1999). 'Resiliency Factors in High School Students at Risk for Academic Failure'. Unpublished doctoral dissertation, California School of Professional Psychology.

111 Moller, J., & Koller, O. (2000). 'Spontaneous and Reactive Attributions Following Academic Achievement'. *Social Psychology of Education*, 4, 67–86.

112 Rebetez, M. M. L., Barsics, C., Rochat, L., D'Argembeau, A., & Van der Linden, M. (2016). 'Procrastination, Consideration of Future Consequences, and Episodic Future Thinking'. *Consciousness and Cognition*', 42, 286–92.

113 Hershfield, H. E., Goldstein, D. G., Sharpe, W. F., Fox, J., Yeykelis, I., Carstensen, I. I., & Bailenson, J. N. (2011). 'Increasing Saving Behavior through Age-Progressed Renderings of the Future Self'. *Journal of Marketing Research*, 48, S23–S37.

114 Interview with author.

115 Tu, Y., & Soman, D. (2014). 'The Categorization of Time and Its Impact on Task Initiation'. *Journal of Consumer Research*, 41(3), 810–22.

116 Brannon, L. A., Hershberger, P. J., & Brock, T. C. (1999). 'Timeless Demonstrations of Parkinson's First Law'. *Psychonomic Bulletin & Review*, 6, 148.

117 Conte, J. M., & Jacobs, R. R. (2003). 'Validity Evidence Linking Poly-chronicity and Big 5 Personality Dimensions to Absence, Lateness, and Supervisory Performance Ratings'. *Human Performance*, 16, 107–29.

118 Ellis, D. A., & Jenkins, R. (2015). 'Watch-Wearing as a Marker of Conscientiousness'. *PeerJ*, 3, e1210.

119 Conte, J. M., Honig, H. H., Dew, A. F., & Romano, D. M. (2001). 'The Incremental Validity of Time Urgency and Other Type A Subcompo-nents in Predicting Behavioral and Health Criteria'. *Journal of Applied Social Psychology*, 31, 1727–48.

120 Pronin, E., Olivola, C. Y., & Kennedy, K. A. (2008). 'Doing Unto Future Selves as You Would Do Unto Others: Psychological Distance and Decision Making'. *Personality and Social Psychology Bulletin*, 34, 224–36.

121 Horn, J., Nelson, C. E., & Brannick, M. T. (2004). 'Integrity, Conscien-tiousness, and Honesty'. *Psychological Reports*, 95(1), 27–38.

122 Dunlop, P., Lee, K., Ashton, M. C., Butcher, S., & Dykstra, A. (2015). 'Please Accept My Sincere and Humble Apologies: The HEXACO Model of Personality and the Proclivity to Apologize'. *Personality and Individual Differences*, 79, 140–5.

123 http://www.svengrahn.pp.se/trackind/jodrell/jodrole2.htm#Zon-d5hide
124 Interview with author
125 Interview with author.
126 Kluger, J. (2017). *Apollo 8: The Thrilling Story of the First Mission to the Moon*. Henry Holt and Company: New York.
127 Ibid.
128 Ibid.
129 Ibid.
130 Ibid.
131 Interview with author.
132 Interview with author.
133 Interview with author.
134 Interview with author.
135 Interview with author.
136 Roth, S., & Cohen, L. J. (1986). 'Approach, Avoidance, and Coping With Stress'. *American Psychologist*, 41(7), 813–19.
137 All the quotes in this section come from an interview with the author.
138 http://www.thehistoryreader.com/contemporary-history/jim-lovell/
139 Jerry Bostick, interview with author.
140 Harland, D. M. (2007). 'The First Men on the Moon: The Story of Apollo 11'. Springs: New York.
141 https://www.popularmechanics.com/space/moon-mars/a4272/4317732/
142 Pfeiffer, C. J. (1965). 'Space Gastroenterology: A Review of the Physiology and Pathology of the Gastrointestinal Tract as Related to Space Flight Conditions'. *Medical Times*, 93, 963–78. Calloway, D. H. & Murphy, E. L. (1969). 'Intestinal Hydrogen and Methane of Men Fed Space Diet'. *Life Science and Space Research* 7, 102–9.

143 Kranz, G. (2000). *Failure Is Not an Option: Mission Control from Mercury to Apollo 13 and Beyond.* Simon and Schuster: New York.

144 Interview with author.

145 Interview with author.

146 https://www.archives.gov/files/presidential-libraries/events/centennials/nixon/images/exhibit/rn100-6-1-2.pdf. The speech was written by William Safire.

147 Interview with author.

148 Interview with author.

149 Interview with author.

150 My thanks to Andrew Baird, from Auburn University, for providing much of the information in this section. For a detailed look at Tindall's work, please see Andrew's great article: (2007) 'How to Land Next to a Surveyor: Bill Tindall and the Apollo Pin-Point Lunar Landing'. *Quest,* 14:2, 19–27. Many of the Tindallgrams are stored in the NASA archive and are available online.

151 Swanson, G. (ed.). (2012). *Before This Decade Is Out: Personal Reflections on the Apollo Program.* Dover Publications: New York. Kranz interview: p. 139.

152 The quotes in this section are taken from NASA's official air to ground transcript: https://www.hq.nasa.gov/alsj/a11/a11transcript_tec.html

153 Interview with author.

154 Interview with author.

155 For an accessible and comprehensive review of this work, see: Fox, E. (2013). *Rainy Brain, Sunny Brain: The New Science of Optimism and Pessimism.* Random House: London.

156 For a review of this work, see: Norem, J. (2002). *The Positive Power of Negative Thinking.* Basic Books: New York.

157 Klein, G. (2007). 'Performing a Project Premortem'. *Harvard Business Review*, 85 (9), 18–19.

158 De Monchaux, N. (2011). *Spacesuit: Fashioning Apollo*. MIT Press: Cambridge, MA. Delaware, F. (2013). 'Apollo Space Suit 1962–1974. A Historic Mechanical Engineering Landmark'. Apollo Space Suit International Latex Corporation.

159 Interview with author.

160 For further information, see: http://www.flippers.be/stern_orbitor_one_history.html

161 Moran, J. (2013). *Armchair Nation: An Intimate History of Britain in Front of the TV*. Profile Books: London.

162 Interview with author.

163 Interview with author.

164 Aldrin, B., & Abraham, K. (2016). *No Dream Is Too High: Life Lessons From a Man Who Walked on the Moon*. National Geographic: Washington DC.

165 Interview with author.

166 https://www.hq.nasa.gov/alsj/a11/a11.launch.html

167 Collins, M. (1975). *Carrying the Fire: An Astronaut's Journey*. W H Allen: London.

168 Nixon, R. (July 24, 1969). Remarks to Apollo 11 Astronauts Aboard the U.S.S. Hornet Following Completion of Their Lunar Mission. For a full text of Nixon's conversation with the astronauts, see http://www.presidency.ucsb.edu/ws/?pid=2138

169 https://www.today.com/news/nev-family-exclusive-we-wouldnt-have-lasted-another-two-days-2D11744232

170 See, for example: Ben-Itzhak S., Bluvstein I., & Maor M. (2014). 'The Psychological Flexibility Questionnaire (PFQ): Development, Reliability and Validity'. Webmed Central PSYCHOLOGY, 5 (4). Fletcher,

B. C. & Stead, B. (2000). *(Inner) FITness and the FIT Corporation (Smart Strategies)*. International Thomson Press: London. Ployhart R. E., & Bliese P. D. (2006). 'Individual Adaptability (I-ADAPT) Theory: Conceptualizing the Antecedents, Consequences, and Measurement of Individual Differences in Adaptability', in *Understanding Adaptability: A Prerequisite for Effective Performance Within Complex Environments*, Vol. 6, eds. Burke C. S., Pierce L. G., & Salas E., Elsevier Science: St. Louis, MO;, 3–39.

171 Bond, F. W., & Bunce, D. (2003). 'The Role of Acceptance and Job Control in Mental Health, Job Satisfaction, and Work Performance'. *Journal of Applied Psychology*, 88, 1057–67.

172 Bond, F. W., & Flaxman, P. E. (2006). 'The Ability of Psychological Flexibility and Job Control to Predict Learning, Job Performance, and Mental Health'. *Journal of Organizational Behavior Management*, 26, 113–30. Ingram, M. P. B. (1998). 'A Study of Transformative Aspects of Career Change Experiences and Implications for Current Models of Career Development'. PhD Dissertation, Texas A & M.

173 'THE FLUX REPORT: Building a resilient workforce in the face of flux'. (2014). Right Management.

174 Fletcher, B. C., Page, N., & Pine, K. J. (2007). 'A New Behavioural Intervention for Tackling Obesity: Do Something Different'. *European Journal of Nutraceuticals and Functional Foods*, 18, 8–9. Fletcher, B. C., Hanson, J., Pine, K. J., & Page, N. (2011). 'FIT – Do Something Different: A New Psychological Intervention Tool for Facilitating Weight Loss'. *Swiss Journal of Psychology*, 70, 25–34. Fletcher, B. C., & Page, N. (2008). 'FIT Science for Weight Loss – a Controlled Study of the Benefits of Enhancing Behavioural Flexibility'. *European Journal of Nutroceuticals & Functional Foods*, 19, 20–3.

175 Fletcher, B. C., & Pine, K. J. (2012). *Flex: Do Something Different.* University of Hertfordshire Press: Hatfield.

176 Ritter, S. M., Damian, R. I., Simonton, D. K., van Baaren, R. B., Strick, M., Derks, J., & Dijksterhuis, A. (2012). 'Diversifying Experiences Enhance Cognitive Flexibility'. *Journal of Experimental Social Psychology*, 48, 961–4.

177 Petrusewicz, M. (2004). 'Note to Entrepreneurs: Meet New People'. *Stanford Report*, 21 January.

178 Fisher, S. G., Macrosson, W. K., & Wong, J. (1998) 'Cognitive Style and Team Role Preference'. *Journal of Managerial Psychology*, 13(8), 544–57.

179 Maddux, W. M., & Galinsky, A. D. (2009). 'Cultural Borders and Mental Barriers: The Relationship Between Living Abroad and Creativity'. *Journal of Personality and Social Psychology*, 96 (5), 1047–61.

180 https://www.virgin.com/richard-branson/meeting-the-dice-man

181 Exline, J., & Hill, P. (2012). 'Humility: A Consistent and Robust Predictor of Generosity'. *Journal of Positive Psychology*, 208–18. Owens, B. P., Johnson, M. D., & Mitchell, T. R. (2013). 'Expressed Humility in Organizations: Implications for Performance, Teams, and Leadership'. *Organization Science*, 24 (5), 1517–38.

APPENDIX

The Astronaut Challenge

The Astronaut Challenge is impossible if you are working on a flat sheet of paper. The mathematical proof of this is somewhat complicated, and involves graph theory, nodes, vertices and the Jordan Curve Theorem. However, to get a rough idea of the issues involved, let's explore the two different scenarios that frequently emerge when you try to solve the puzzle.

In one scenario you connect two of the astronauts to all three tanks and you end up with the third astronaut on the 'outside' of the lines like this:

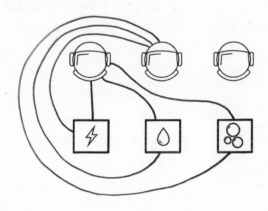

As you can see, it's now impossible to connect the third astronaut to the electricity supply without crossing a line.

In a second scenario, the third astronaut ends up 'inside' the lines like this:

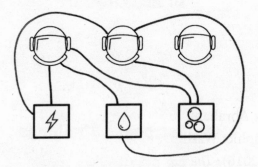

And you can see that it will be impossible to connect the astronaut to the electricity.

The most interesting way of solving the puzzle involves changing the rules and working on a curved surface. If you want to explore this approach, grab a bagel and a pen and mark the three astronauts around the top surface of the bagel. Next, turn the bagel over and mark the three tanks on the bottom of the bagel. Now see if you can connect the three astronauts with the three tanks without any of the lines crossing one another. Under these circumstances it really is possible!

BIBLIOGRAPHY

Aldrin, B. & Abraham, K. (2016). *No Dream Is Too High*. National Geographic: Washington DC.

Bostick, J. (2016). *The Kid from Golden. From the Cotton Fields of Mississippi to NASA Mission Control and Beyond*. iUniverse: Bloomington, IN.

Chaikin, A. (1994). *A Man on the Moon*. Penguin Books: New York.

De Monchaux, N. (2011). *Spacesuit: Fashioning Apollo*. MIT Press: Cambridge, MA.

Farmer, G. & Hamblin, D. J. (1970). *First on the Moon: A Voyage with Neil Armstrong, Michael Collins and Edwin E. Aldrin, Jr*. Little Brown: Boston, MA.

Hadfield, C. (2015). *An Astronaut's Guide to Life on Earth*. Pan-Macmillan: London.

Hansen, J. R. (1995). *Enchanted Rendezvous: John C. Houbolt and the Genesis of the Lunar-Orbit Rendezvous Concept. Monograph in Aerospace History*, No. 4, NASA History Division: Washington DC.

Hill, P. S. (2018). *Mission Control Management: The Principles of High Performance and Perfect Decision-Making Learned from Leading at NASA*. Nicholas Brealey Publishing: Boston, MA.

Houston, R. & Heflin, M. (2015). *Go, Flight! The Unsung Heroes of Mission Control*. University of Nebraska Press: Lincoln, NB.

Kluger, J. (2017). *Apollo 8: The Thrilling Story of the First Mission to the Moon*. Henry Holt and Company: New York.

Kraft, C. (2001). *Flight: My Life in Mission Control*. Dutton Books: New York.

Kranz, G. (2009). *Failure Is Not an Option: Mission Control from Mercury to Apollo 13 and Beyond*. Simon & Schuster: New York.

Lunney, G. S., Bostick, J., Reed, H. D., Deiterich, C. F., Bales, S. G., Gravett, W., Kennedy, M., von Ehrenfried, M., Boone, W. J., & Stoval, W. (2012). *From The TRENCH of Mission Control to the Craters of the Moon*. CreateSpace Independent Publishing Platform: CA.

Maher, N. M. (2017). *Apollo in the Age of Aquarius*. Harvard University Press: Cambridge, MA.

Murray, C. A. & Cox, C. B. (1989). *Apollo: The Race to the Moon*. Simon & Schuster: New York.

Neufeld, M. J. (2013). *The Rocket and the Reich: Peenemunde and the Coming of the Ballistic Missile Era*. Smithsonian Books: DC.

Potter, C. (2017). *The Earth Gazers*. Head of Zeus, Ltd: London.

Scott, D. M. & Jurek, R. (2014). *Marketing the Moon*. MIT Press: Cambridge, MA.

Swanson, G. (ed.). (2012). *Before This Decade Is Out: Personal Reflections on the Apollo Program*. Dover Publications: New York.

von Ehrenfried, M. (2016). *The Birth of NASA*. Springer: New York.

von Ehrenfried, M. (2018). *Apollo Mission Control: The Making of a National Historic Landmark*. Springer: New York.

Ward, R. (2009). *Dr. Space: The Life of Wernher Von Braun*. Naval Institute Press: MD.

ACKNOWLEDGEMENTS

This book wouldn't have happened without the assistance and support of lots of people.

First, I originally thought of the idea while I was chatting with comedian and space fan Helen Keen. Helen has been tremendously supportive and helpful throughout and kindly put me in touch with another space fan, Craig Scott. Helen, thank you.

Craig is, quite frankly, amazing. He is passionate about space exploration and has managed to become firm friends with many of the mission controllers. Craig was kind enough to introduce me to this remarkable group of people, was enormously supportive during the early stages of the project, and continued to provide invaluable assistance along the way. Craig, I am so grateful for your time, energy and goodwill. Thank you.

Then we come to the mission controllers themselves. A huge thank you to all the people who gave up their valuable time to email and chat: Steve Bales, Jerry Bostick, Charles Deiterich, Manfred von Ehrenfried, Ed Fendell, Gerry Griffin, Jay Honeycutt, Dick Koos, Glynn Lunney and Doug Ward. Emailing and chatting to you was as pleasurable as it was helpful. It was an honour and a privilege to make your acquaintance. Once again, thank you.

Also, my thanks to top lawyer Terry O'Rourke and wonderful Mission Evaluation Room engineer Jerry Woodfill, who both spoke so passionately about seeing Kennedy's Moon speech at Rice University. I also really appreciated inventor and engineer Dixie Rinehart finding the time to chat, his brother, Palfi, for making it all happen and being such good fun, and daughter Tanya for being so helpful. I am also grateful to Andrew Baird from Auburn University for providing such fascinating and helpful insights into the strange world of Tindallgrams. I would also like to thank that passionate lover of space exploration 'Ozzie' Osband for corresponding about telephone numbers and rocket launches. And a special thanks to the amazing James Burke – chatting to you was a total pleasure and yes, I did remember to give the book a title!

Also, many thanks to David Britland, Ken Gilhooly, Colin Uttley and Jeff Sanford for their invaluable encouragement and advice along the way.

This book wouldn't have happened without my brilliant agent Patrick Walsh, and fabulous editors, Jon Butler, Katy Follain and Marian Lizzi.

And finally, as ever, this book simply couldn't have come into being without the wonderful support of my partner, Caroline Watt.

Everyone was amazingly helpful, but when it comes to the facts, figures and psychology, the buck rests firmly with me.

AUTHOR BIOGRAPHY

Professor Richard Wiseman has been described by one *Scientific American* columnist as 'the most interesting and innovative experimental psychologist in the world today'.

A passionate advocate for science, Richard has written several best-selling books, including *The Luck Factor* (examining the lives and minds of lucky people), *Quirkology* (exploring the curious science of everyday life), and *59 Seconds* (investigating the fast-acting psychological techniques that make people happier and more productive).

Richard frequently appears on the media and gives talks. His YouTube videos have received over 500 million views. He was listed in the *Independent on Sunday*'s top 100 people who make Britain a better place to live, and currently holds Britain's only Professorship in the Public Understanding of Psychology at the University of Hertfordshire.

Richard's research has been published in some of the world's leading academic journals and he has delivered keynote addresses to organizations across the world, including The Swiss Economic Forum, Google and Amazon.

273